BrightRED Study Guide

CfE HIGHER

GEOGRAPHY

Lynn Cockburn and Valerie Nicol

First published in 2015 by:
Bright Red Publishing Ltd
1 Torphichen Street
Edinburgh
EH3 8HX

A CIP record for this book is available from the British Library.

ISBN 978-1-906736-63-7

With thanks to:
PDQ Digital Media Solutions Ltd, Bungay (layout) and Anne Horscroft (copy-edit).

Cover design and series book design by Caleb Rutherford – e i d e t i c.

Acknowledgements

Every effort has been made to seek all copyright-holders. If any have been overlooked, then Bright Red Publishing will be delighted to make the necessary arrangements.

Permission has been sought from all relevant copyright holders and Bright Red Publishing is grateful for the use of the following:

Chris Combe (CC BY 2.0)[1] (p 10); Roy Luck (CC BY 2.0)[1] (p 10); Nick Bramhall (CC BY-SA 2.0)[2] (p 10); Mike63/iStock.com (p 11); David Goddard/Getty Images News/Getty Images (p 11); Jynto (public domain) (p 11); Boschfoto (CC BY-SA 3.0)[3] (p 12); 6 images by Christopher Lofthouse (pp 12, 13, 14, 15, 89 & 93); Iulus Ascanius (public domain) (p 12); Lies Thru a Lens (CC BY 2.0)[1] (p 15); swifant (CC BY-SA 2.0)[2] (p 15); Leon (CC BY-ND 2.0)[4] (p 16); Jim Linwood (CC BY 2.0)[1] (p 16); Graeme Churchard (CC BY 2.0)[1] (p 17); Ordnance Survey Maps © Crown Copyright. All rights reserved. Licence number 100049324 (pp 13 & 17); Anjo Kan/iStock.com (p 21); Les Chatfield (CC BY 2.0)[1] (p 22); Storye book (CC BY 3.0)[5] (p 22); Martin Robson (CC BY-SA 2.0)[2] (p 22); Moorefam/iStock.com (p 22); tothemoonphoto/iStock.com (p 27); Extract from 'National 4&5 Geography: Human Environments' by Calvin and Suzi Clarke published by Hodder Gibson 2013 (p 40); © ONS. Reproduced with permission of ONS Census Transformation Programme (p 42); BartekSzewczyk/iStock.com (p 43); michaboon88/iStock.com (p 45); Poprotskiy Alexey/Shutterstock.com (p 46); fpolat69/Shutterstock.com (p 48); European Union, 1995-2015/ECHO (p 48); Raymond McCrae (CC BY 2.0)[1] (p 52); Brendan Howard/Shutterstock.com (p 53); Daniel (CC BY 2.0)[1] (p 54); Glasgow City Council – www.glasgow.gov.uk (p 55); Paul Walter (CC BY 2.0)[1] (p 56); Martin McCarthy/iStock.com (p 56); moonmeister/iStock.com (p 56); Adisa/Shutterstock.com (p 58); Prefeitura/EOM (p 60); Mario Roberto Duran Ortiz Mariordo (CC BY 3.0)[5] (p 61); Ana Paula Hirama (CC BY 2.0)[1] (p 61); Leandro Neumann Ciuffo (CC BY 2.0)[1] (p 62); chensiyuan (CC BY-SA 4.0)[6] (p 65); Andréa Farias Farias (CC BY-SA 2.0)[2] (p 65); Ricardo Stuckert (CC BY 3.0 BR)[7] (p 66); Philippe Rekacewicz, UNEP/GRID-Arendal (Source: UNEP, International Soil Reference and Information Centre (ISRIC), World Atlas of Desertification, 1997) (p 68); Flockedereisbaer (CC BY-SA 3.0 DE)[8] (p 70); Benedikt.Seidl (public domain) (p 70); Michael Jansen (CC BY-ND 2.0)[4] (p 70); A.Davey (CC BY 2.0)[1] (p 72); dorothy.voorhees (CC BY-SA 2.0)[2] (p 73); Sustainable Sanitation Alliance (CC BY 2.0)[1] (p 73); Djpalmer93 (CC BY-SA 4.0)[6] (p 75); Southbank Centre (CC BY 2.0)[1] (p 76); Radoslaw Botev (CC BY-SA 3.0)[3] (p 77); Calin Tatu/Shutterstock.com (p 78); Department of Foreign Affairs and Trade (CC BY 2.0)[1] (p 79); James Gathany/CDC (public domain) (p 82); Petaholmes (public domain) (p 82); Alison Bird/USAID (public domain) (p 82); Africa Studio/Shutterstock.com (p 84); yangna/iStock.com (p 85); Simon Berry (CC BY-SA 2.0)[2] (p 87); fridgeirsson (CC BY-ND 2.0)[4] (p 89); hroe/iStock.com (p 89); hamster (CC BY-ND 2.0)[4] (p 90); Scott Barron – non-commercial (p 90); NOAA/NASA GOES Project (CC BY 2.0)[1] (p 93); Charles H. Smith (public domain) (p 93); Ola Matsson (CC BY 2.0)[1] (p 94); Oiva Eskola (CC BY 2.0)[1] (p 94); Sustainable Sanitation Alliance (CC BY 2.0)[1] (p 94); HPA, Laos, 2014 (CC BY-ND 2.0)[4] (p 95); Exam questions © Scottish Qualifications Authority (n.b. solutions do not emanate from the SQA) (pp 13, 17, 24, 33 & 91).

(CC BY 2.0)[1]   http://creativecommons.org/licenses/by/2.0/
(CC BY-SA 2.0)[2]   http://creativecommons.org/licenses/by-sa/2.0/
(CC BY-SA 3.0)[3]   http://creativecommons.org/licenses/by-sa/3.0/
(CC BY-ND 2.0)[4]   http://creativecommons.org/licenses/by-nd/2.0/
(CC BY 3.0)[5]   http://creativecommons.org/licenses/by/3.0/
(CC BY-SA 4.0)[6]   https://creativecommons.org/licenses/by-sa/4.0/
(CC BY 3.0 BR)[7]   http://creativecommons.org/licenses/by/3.0/br/
(CC BY-SA 3.0 DE)[8] http://creativecommons.org/licenses/by-sa/3.0/de/

Printed and bound in the UK by Martins the Printers.

INTRODUCTION

COURSE SUMMARY

This is a revision guide and will help to supplement your course notes from school in preparation for the CfE Higher Geography exam. You may have studied CfE National 5 Geography last year and you will be familiar with some of the topics. This year you will look at these topics in greater depth and will study some new ones.

COURSE OVERVIEW

CfE Higher Geography has the following course units:

Physical Environments	Human Environments	Global Issues
Biosphere	Population	Development and Health
Lithosphere	Urban	Global Climate Change
Hydrosphere	Rural	River Basin Management
Atmosphere		Trade, Aid and Geopolitics
		Energy

You will be assessed at school, either at the end of each unit or throughout the units, to see if you meet the assessment criteria. This will be on a pass/fail basis of a minimum standard.

You will also be asked to submit an assignment – we will consider this later – as part of the coursework, and in the exam you will be asked to use a wide range of geographical methods and techniques, including the interpretation of Ordnance Survey maps, tables and graphs.

THE EXAM – 60 MARKS

The exam lasts for 2 hours 15 minutes and is made up of four sections, as outlined in the following table.

Section	Topics covered and type of question	Number of questions to be attempted	Marks allocated (60 marks total)
Section 1	**Physical Environments** Extended response questions of 4–6 marks each	Attempt all questions	15 marks
Section 2	**Human Environments** Extended response questions of 4–6 marks each	Attempt all questions	15 marks
Section 3	**Global Issues** Extended response questions of 4–6 marks each	Attempt two out of five questions	10 marks for each question
Section 4	**Application of Geographical Skills** Extended response question requiring the application of geographical skills acquired during the course. The skills assessed in the exam will include mapping skills and the use of numerical/graphical information	Attempt question	10 marks

Roughly speaking, you should allow 30 minutes for each section in the exam. This will allow reading time at the beginning and end of the paper. Each section requires application of the knowledge you have gained in class throughout the year. You will also have to show your understanding of skills acquired throughout the year in the assignment.

DON'T FORGET

You need to **develop** points in the exam. Lists will not gain much credit – a maximum of one mark – but well-annotated diagrams along with summaries could gain full marks. You will need to practise doing questions under timed conditions. The last thing you want to do is run out of time!

ASSIGNMENT – 30 MARKS

Write up: 1 hour 30 minutes

The geography assignment will provide you with the opportunity to do some research into a geographical topic or issue. Your teacher will encourage you to use a range of fieldwork techniques to gather information, process this information and then write up the findings.

The assignment will give you the opportunity to demonstrate the following higher order cognitive skills, knowledge and understanding:

- identifying a geographical topic or issue

- carrying out research, which should include fieldwork where appropriate

- demonstrating knowledge of the suitability of the methods and/or reliability of the sources used

- processing and using a range of information gathered

- drawing on detailed knowledge and understanding of the topic or issue

- analysing information from a range of sources

- reaching a conclusion supported by a range of evidence on a geographical topic or issue

- communicating information

The assignment will be carried out at some point during the year and will be written up under assessment conditions. You will have to communicate your findings in a form that shows evidence of the skills you have used and your knowledge and understanding of the geographical issue or topic studied. The aim of the presentation of findings is to assess the quality of your research into, and analysis, knowledge and understanding of the issue or topic.

This is marked by an examiner and will contribute to your overall mark for the exam.

 **DON'T FORGET**

The more organised you are, the better your assignment is likely to be!

HOW THIS GUIDE CAN HELP YOU

There is no shortcut to passing any course at Higher level. To obtain a good pass requires consistent, regular revision over the duration of the course. The aim of this study guide is to help you to achieve success by providing you with concise and engaging coverage of the CfE Higher Geography course material. We recommend that you use this book in conjunction with your class notes to revise each topic area, prepare for unit assessments and the assignment, and in your preparation for the final exam.

SOILS

SOIL: THE FACTS

- Without soil, life on Earth would not exist.

- Soil is our most valuable non-renewable resource and provides all of the land-based food supply for plants, humans and animals.

- Soils are the link between the Biosphere, Lithosphere, Hydrosphere and Atmosphere units of Physical Environments in the CfE Higher Geography course.

FACTORS AFFECTING SOIL FORMATION

Main factors affecting soil formation.

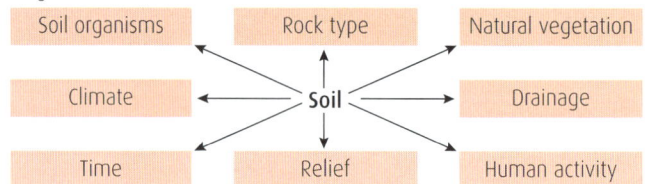

Rock type	The parent material/regolith influences the soil. Harder rocks often produce thin soils, whereas softer rocks weather more quickly. Different rock types can produce either acidic or alkaline soils and influence the colour of the soil. Soil texture is also affected by rock type and this is reflected in the permeability and composition of the soil.
Natural vegetation	Soil humus is affected by the natural vegetation litter. Coniferous forests produce acidic (mor) humus, whereas deciduous forests form chemically neutral soils (mull) with a rich soil fauna.
Drainage	Poorly drained soils encourage peaty soils to form (gley), usually in areas where there are gentle slopes.
Human activity	In Scotland, ancient forests were cut down by prehistoric people. This caused soil erosion and the development of heather moorlands.
Relief	Steep slopes may limit soil formation through mass movement and organic decomposition is slower at high altitudes.
Time	It may take at least 100 years to form 1 cm of soil. Soils in the UK take longer to form due to latitude. They are also relatively young and are therefore not very deep. This is mainly a result of the effects of glaciers on the land during the ice ages.
Climate	Soils take longer to form in colder areas as organic decomposition is slower; warmer areas allow faster decomposition and breakdown of organic material. Wetter areas encourage leaching. Eluviation occurs when soluble minerals and humus are washed out of the A horizon; illuviation occurs when this material is washed into the B horizon.
Soil organisms	Creatures such as insects and worms help to mix soils, encourage air into the soil and add their waste products to the soil chemistry. They also affect the breakdown and decay of vegetation in the humus layer.

Soil is made up of mineral matter from:

- rock/regolith broken down by weathering

- organic matter from decaying plants, helped to decompose by the soil biota

- dead organisms broken down by other organisms such as earthworms

- air and water – some soils have more or less of these; well-drained soils have more air than waterlogged soils.

SOIL FORMATION

Humification is the process by which leaf litter is broken down into humus/organic material.

Leaching is the movement of minerals and humus downwards through a soil in areas where precipitation is greater than evaporation.

Capillary action is when nutrients are drawn upwards through a soil and occurs when evaporation is greater than precipitation.

THE SOIL SYSTEM

Some of the main inputs, processes and outputs of the soil system are shown in the following diagram.

Inputs:
- Water from the atmosphere
- Nutrients from decaying rocks
- Excretions from plant roots
- Solar energy and gases
- Organic matter from decaying plants and organisms

Processes:
- Physical weathering
- Chemical weathering
- Leaching
- Gleying
- Capillary action
- Mixing and incorporation of decaying vegetation by soil fauna

Outputs:
- Loss of soil by soil creep
- Nutrients taken up by plants
- Nutrients lost in water passing through soil
- Evaporation: loss of water from topsoil
- Water and nutrient loss by capillary action

The soil system.

ONLINE

Head to www.brightredbooks.net to investigate soils further.

CLASSIFICATION OF SOIL PROFILES

Soils are divided into a series of different layers known as **horizons**.

The A_0 **horizon** is the uppermost layer, consisting of organic matter in various stages of decomposition:

- the **L** or **leaf litter** layer consists of undecomposed pine cones and needles, grass and leaves that have fallen on top of the soil

- the **F** or **fermentation** layer shows the first stages of decomposition

- the **H** or **humus** layer consists of completely decomposed plant and animal matter.

The main nutrient-rich layer of **topsoil** is known as the **A horizon** and consists of a mixture of mineral matter from the weathered parent material and organic matter introduced from above.

The **B horizon**, or **subsoil**, has a coarser texture and contains more mineral matter from the weathered parent material. Soluble organic matter may also be washed down from above by **leaching**.

The **C horizon** is the regolith (weathered parent material), which contributes mineral matter to the upper horizons.

The **E horizon**, found in **podzol** profiles, is ash grey in colour and forms an iron pan layer. Minerals collect in this layer through **illuviation** and **eluviation**. Illuviation is leaching – when minerals are washed into or migrate from the topsoil into the next horizon. Eluviation is the upwards movement of minerals from the parent rock.

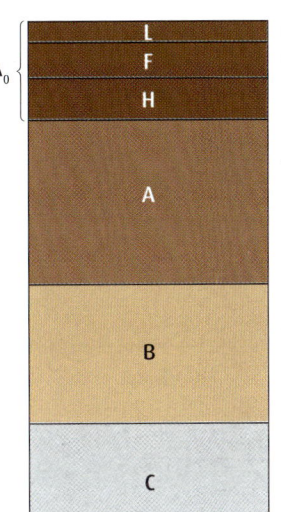

L	Leaf litter
F	Decomposing layer
H	Humus layer
A	Topsoil (organic and mineral matter)
B	Subsoil (little organic matter)
C	Weathered parent material (regolith)

A_0 is marked against the L, F, and H layers.

A model soil profile.

THINGS TO DO AND THINK ABOUT

Use mind-maps to test yourself on the factors involved in soil formation and practise drawing the soil profile. Both will help you to visualise these topics for the exam to produce detailed descriptions.

ONLINE TEST

Head to www.brightredbooks.net to test yourself on this topic.

SOIL PROFILES

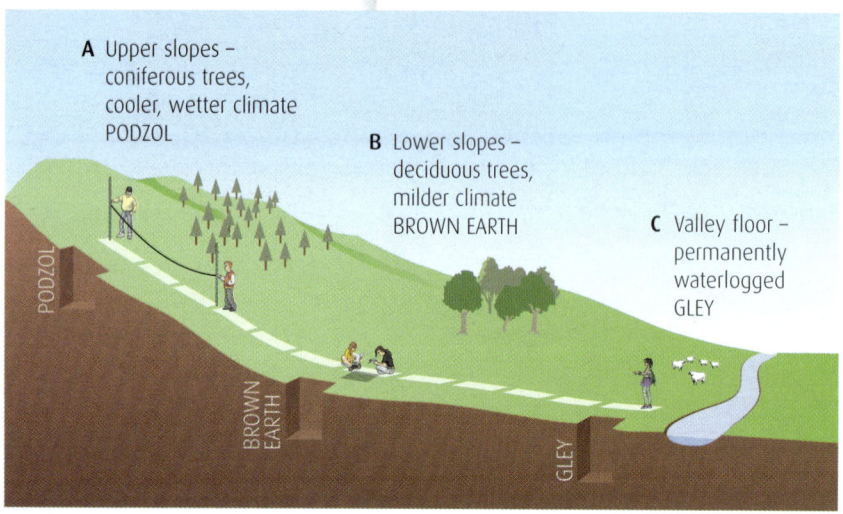

A Upper slopes – coniferous trees, cooler, wetter climate PODZOL

B Lower slopes – deciduous trees, milder climate BROWN EARTH

C Valley floor – permanently waterlogged GLEY

PODZOL

BROWN EARTH

GLEY

A soil catena showing the soil types that develop down a slope.

SOIL CATENAS

A soil catena is a sequence of soil types that develop down a slope. In general, gleys are found at the bottom of the slope where water collects, brown earths are found slightly higher up where the temperature is mild, and podzols are found near the top of the slope where temperatures are cooler.

When describing each soil type, you need to mention:

- natural vegetation
- soil organisms
- climate and relief
- horizons and colours
- soil uses.

SOIL TYPES

Coniferous forest, short cool summers

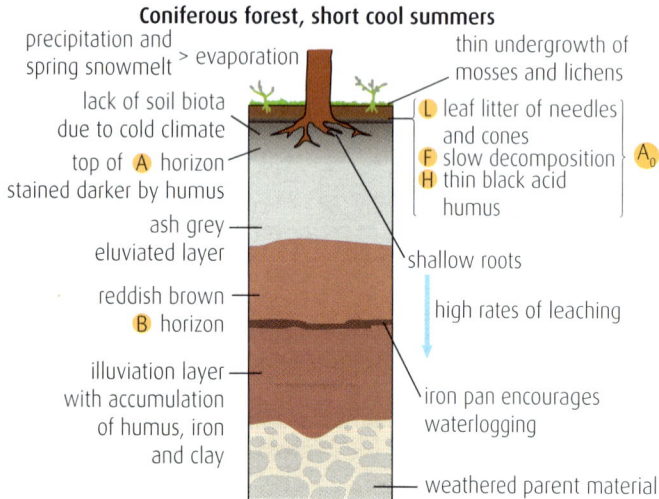

precipitation and spring snowmelt > evaporation

lack of soil biota due to cold climate

top of **A** horizon stained darker by humus

ash grey eluviated layer

reddish brown **B** horizon

illuviation layer with accumulation of humus, iron and clay

thin undergrowth of mosses and lichens

L leaf litter of needles and cones
F slow decomposition
H thin black acid humus
A_0

shallow roots

high rates of leaching

iron pan encourages waterlogging

weathered parent material

Profile of a podzol.

Mild climate, moderate rainfall

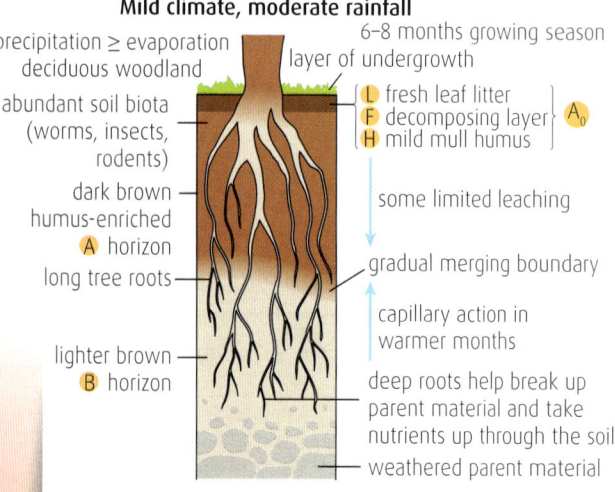

precipitation ≥ evaporation deciduous woodland

abundant soil biota (worms, insects, rodents)

dark brown humus-enriched **A** horizon

long tree roots

lighter brown **B** horizon

6–8 months growing season
layer of undergrowth

L fresh leaf litter
F decomposing layer
H mild mull humus
A_0

some limited leaching

gradual merging boundary

capillary action in warmer months

deep roots help break up parent material and take nutrients up through the soil

weathered parent material

Profile of a brown earth soil.

Podzols

Podzols are found in areas of coniferous forest. The leaf litter of pine needles is difficult to break down because winters are cold and summers cool on the higher slopes. The colder climate means there are very few soil organisms to disturb the soil, resulting in more distinct soil horizons. Moisture moves downwards through the soil, leaching minerals from the A horizon and depositing them in the B horizon. The A horizon is grey due to a lack of organic material and the B horizon is reddish brown in colour. The A horizon is sandy in texture and the B horizon has a denser texture. An iron pan may develop, impeding drainage and causing waterlogging. The rock type in these areas may be fluvio-glacial sands, till or an acidic parent rock. The mor humus is thin and acidic, making podzols fairly infertile. They are mostly used for commercial forestry or rough grazing.

Brown earth soils

Brown earth soils are found in areas of deciduous woodland. The leaf litter breaks down each autumn, producing a deep soil. Tree roots and soil organisms (earthworms and moles) mix the soil, resulting in very little difference in colour between the horizons. The mild climate aids rapid decomposition of the leaf litter and a thick mull humus develops. Leaching is low as evaporation exceeds precipitation and capillary action may cause minerals to rise through the soil profile during the summer. The soil colour varies from black humus to dark brown in the A horizon to lighter brown in the B horizon, where the humus content is less obvious. The soil texture is loamy and well-aerated in the A horizon, but lighter in the B horizon. These are the most agriculturally productive of the three soils described here as they tend to be deeper and more fertile.

contd

Gley soils

The natural vegetation in gley soils comes from shrubs, grasses and rushes. The decay of leaf litter is slow because these areas are permanently or temporarily waterlogged as a result of the cooler/wetter climate. The soils lose oxygen because the voids are filled with water. This creates anaerobic conditions and there are very few soil fauna. Waterlogging causes the iron in the soil to change colour from red/brown to blue/grey. This creates a distinctive mottled appearance in the B horizon. The A horizon tends to be dark brown in colour. The humus layer is very thin and almost black. Gley soils are found at the bottom of slopes where water accumulates. They require drainage before rough grazing or forestry can take place.

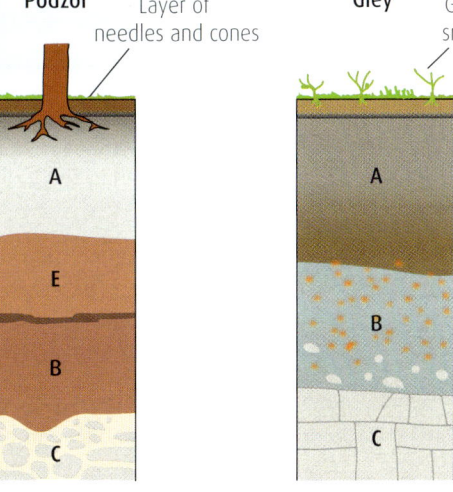

Long winters
precipitation > evaporation
gentle relief
hinders drainage
anaerobic conditions and little soil biota
waterlogged blue/grey **B** horizon

vegetation of shrubs, grasses and rushes
mosses and lichens on surface
slowly decomposing leaf litter
thin black acid humus
dark brown/grey colour
build up of organic material in **A** horizon – peat may form
distinct boundary between layers
red/orange mottling produced by iron compounds
rock fragments
impermeable parent material

Profile of a gley soil.

 **ACTIVITY**

Look over each of the soil profiles. Then test yourself – describe each of the soils, then explain why each soil has such different characteristics. You will need to include information about the climate, soil biota, colour and horizons for a good answer.

THINGS TO DO AND THINK ABOUT

1   Look at the diagram below, which shows two soil profiles.

Choose one of the profiles.

(a) Describe in detail, the characteristics of the soil, including horizons, colour, texture and drainage.   (4 marks)

(b) Study the diagram below. Explain in detail how the major soil-forming factors shown in the diagram have contributed to the formation of your chosen soil profile.   (6 marks)

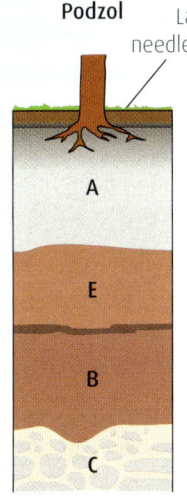

Podzol Layer of needles and cones

A

E

B

C

Gley Grasses and small shrubs

A

B

C

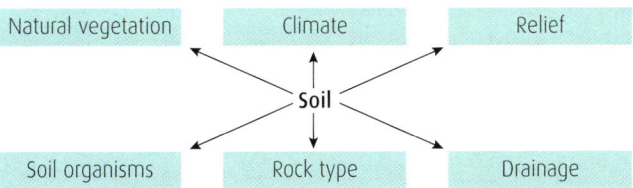

Natural vegetation   Climate   Relief

Soil

Soil organisms   Rock type   Drainage

For the Biosphere topic in the exam, you need to know the properties of and formation processes for:
● podzols   ● brown earth soils   ● gley soils.

GLACIAL EROSION

This chapter focuses on the lithosphere and the features of the physical surface of the Earth. For the exam you will need to know about glacial features and coastal features – both erosional and depositional. You will have covered case studies in class time, some of which may be mentioned in the following sections.

The processes of glacial erosion and deposition occurred over a period of around 2–4 million years. The last ice disappeared about 8000 years ago. Ice created very distinctive landscapes in both upland and lowland areas. Good examples are found in Scotland, such as in the Cairngorms and around Loch Lomond, and also in the English Lake District.

FEATURES OF GLACIAL EROSION

Corries

Frost shattering loosens material

Snow falls in a hollow on a north-facing slope

Surface thaw in summer

Rock debris removed by summer meltwater

Snow compressed to form névé and firn

Moraine from frost shattering

Bergschrund/crevasses

Abrasion, where rocks in the glacier scrape and erode the bedrock, deepens the base

Formation of rock lip

Summer meltwater

Steepening of corrie backwall by plucking – frozen ice pulling off pieces of bedrock

Rotational movement

Moraine on top of, within, beneath and at the end of the glacier

A corrie is an armchair-shaped hollow found near the top of a mountain.

Before the ice formed, there was a shallow hollow at the top of the mountain (north-facing). During the ice age, snow collected in the hollow and did not melt. Gradually, as layers of snow built up, oxygen was squeezed out, resulting in **névé**.

Over time, the weight of the ice and the action of gravity caused the corrie glacier to move by rotational sliding. Plucking (frozen ice pulling rock from the back wall) steepened the corrie. The corrie floor was deepened by abrasion from the rocks and ice in the glacier. The glacier had less power at its front, so rock debris was deposited here to form a lip.

After the ice melts, corries often fill up with a tarn or lochan, such as Red Tarn in the English Lake District and Coire an Lochan in the Cairngorms.

U-shaped valleys

At the peak of the ice age, corrie glaciers extended beyond their initial hollows and flowed into V-shaped valleys, transforming their shape by plucking and abrasion into steep-sided U-shaped valleys (or glacial troughs), such as the Lairig Ghru in the Cairngorms. Misfit streams (rivers too small for the U-shaped valley) are often found on the valley floor after glaciers have retreated. A terminal moraine stretching across the wide floor may cause a ribbon lake to form.

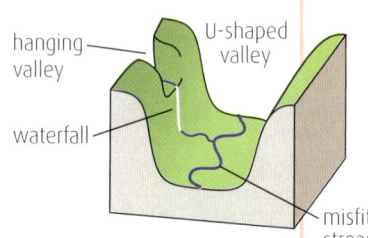

hanging valley

U-shaped valley

waterfall

misfit stream

A U-shaped valley is a steep, deep and wide valley formed by a glacier.

Smaller tributary glaciers with less erosive power also create U-shaped valleys. When the ice melts, these tributary valleys are left 'hanging' above the main valley.

Glencoe, u-shaped valley.

A hanging valley in Alaska.

Loch Avon in the Cairngorms National Park is a ribbon lake with a truncated spur.

contd

Arêtes

When two corries form back-to-back, an arête forms as the backwalls of the corrie meet. Freeze–thaw processes may steepen this ridge. A good example of an arête is Striding Edge in the English Lake District.

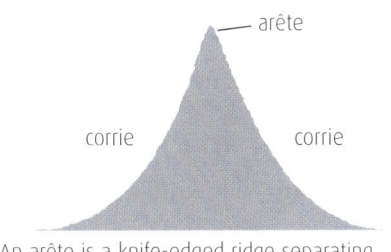

An arête is a knife-edged ridge separating two corries.

The top of Helvellyn, English Lake District.

Pyramidal peaks

If three or more corries form at the top of a mountain, this creates a pyramidal peak such as Ben Lui in Scotland or Helvellyn in the English Lake District.

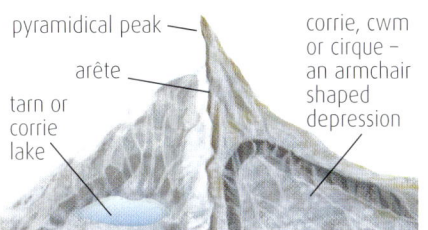

A pyramidal peak is a horn-shaped peak at the top of a mountain.

Crags and tails

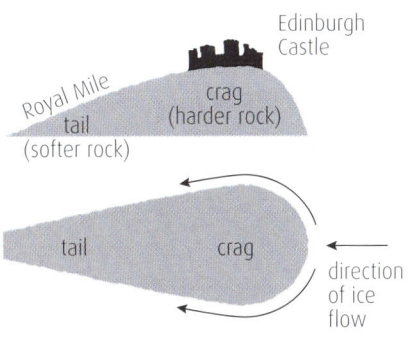

Edinburgh Castle and the Royal Mile are formed from a crag and tail structure.

A good example of a crag and tail structure is Edinburgh Castle rock and the Royal Mile.

If a glacier meets an outcrop of rock that is harder than the rock of the surrounding area, then the sides of the softer rock are eroded by the glacier in the direction of ice flow and the land behind the crag of harder rock is protected. This forms a gently sloping ridge called a tail.

Roche moutonées

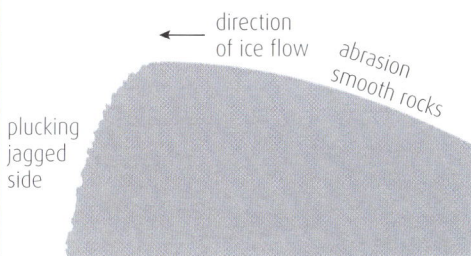

A roche moutonée is an outcrop of smoothed, hard rock.

A roche moutonée in Snowdonia, North Wales.

A roche moutonée is an outcrop of harder rock that was smoothed on the side facing the ice, giving a gentle slope, and plucked on the side facing away from the ice to give a steep slope.

 THINGS TO DO AND THINK ABOUT

Test yourself with mind-maps. Draw a mind-map based on the one shown here and fill in all of the features you can remember.

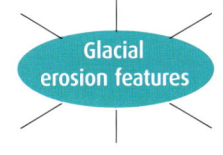

Glacial erosion features

GLACIAL DEPOSITION

Glacial deposition mainly took place in lowland areas. The depositional features seen today in the UK were formed either under the ice or on outwash plains beyond the terminal moraines.

PROCESSES OF GLACIAL DEPOSITION

Glacial **transportation** can move eroded material over long distances. Evidence for this exists in the form of **erratics** – large rocks found a long way from their source area. A mix of rocks, clays, sands and silts is added to moving glaciers by weathering and erosional processes. When the glacier melts, all this **englacial** material is deposited as one of two types of glacial **drift**:

- **unsorted** and **unstratified** material deposited directly by ice is called **till**

- **sorted** and **stratified** material deposited by the **meltwater** flowing out of the glacier forms **fluvio-glacial deposits**

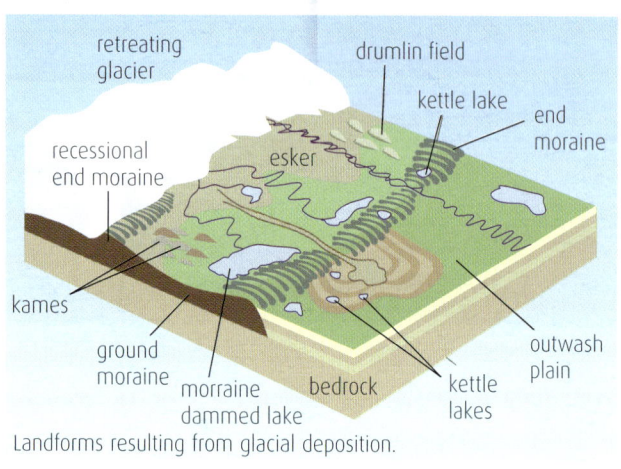

Landforms resulting from glacial deposition.

FEATURES OF GLACIAL DEPOSITION

Drumlins

During the ice ages, clay containing angular rocks covered the lowlands underneath the ice cover. These deposits are now known as **boulder clay** or **till**. The surface of the boulder clay can sometimes be marked by long rounded hills called **drumlins**. Often over 100 m high, drumlins are created when material is deposited as a result of the friction between the ice and the underlying rock, which causes the glacier to deposit its load.

Drumlin formed when a glacier deposited its load of transported material.

Erratics

Glaciers sometimes pick up large boulders and transport them for hundreds of miles. When the glaciers retreat in warmer periods, these rocks are deposited in areas where the surrounding rocks may be of different types.

Erratics, north-west Scotland.

Eskers

Eskers are created inside the ice cover by rivers and streams. These carry stratified sands and gravels and, when the ice retreats, the rivers build up ridge-like deposits.

Eskers, south-east Scotland.

Moraines

The **terminal moraine** marks the maximum extent of a glacier across a valley. **Lateral moraines** are found at the sides of a U-shaped valley and are caused by the freeze–thaw action of bare rocks above the glacier. **Medial moraines** are found in the middle of a U-shaped valley running parallel with the valley sides. **Recessional moraines** are found where the ice retreated, then advanced a little, deposited some material, retreated further, and then advanced again, depositing more material parallel to the terminal moraine.

Terminal moraine marking maximum extent of a glacier.

contd

Kettle holes

Kettle holes are formed when large pieces of ice take some time to melt once the glacier has retreated, causing a depression in the ground that can sometimes be filled by water.

Outwash plains

The **outwash plain** is the area beyond the terminal moraine. The deposits here have been carried away from the glacier by meltwater. They are well-sorted, i.e. the larger materials are closer to the moraine and the finer materials, such as sand, are deposited much further away.

Glacial outwash plain, southern Iceland

Kettle hole, south-east Scotland.

Kames

Kames are irregularly shaped mounds of sands and gravels, generally found on the sides of valleys. They were created by streams running along the side of the ice.

 DON'T FORGET

Learn the names of specific features – you will get credit for these in the exam!

GLACIAL FEATURES ON ORDNANCE SURVEY MAPS

The map shows the main glacial features that can be identified on Ordnance Survey maps. Other features to look for include pyramidal peaks, misfit streams and waterfalls from hanging valleys. Truncated spurs may be identifiable between hanging valleys. Corries predominantly face north.

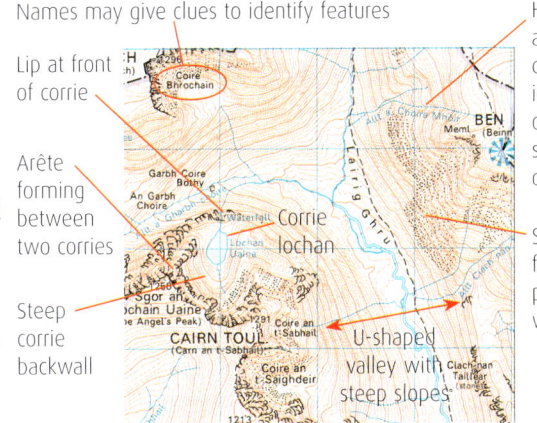

Names may give clues to identify features

Lip at front of corrie

Arête forming between two corries

Steep corrie backwall

Corrie lochan

Hanging valleys are often difficult to identify but may occur where streams flow out of corries

Scree slopes formed by physical weathering

U-shaped valley with steep slopes

Glacial features on maps.

 ONLINE TEST

Head to www.brightredbooks.net and test yourself on this topic.

 ONLINE

For more on glacial deposition, follow the link at www.brightredbooks.net

 DON'T FORGET

Practise the diagrams. The more you practise, the less time will be spent trying to create 'works of art' in the exam. A well-annotated diagram could be awarded full marks in the final exam.

Till deposits are unsorted and unstratified materials deposited by ice and include:

- drumlins
- medial moraines
- terminal moraines
- recessional moraines
- lateral moraines

Fluvio-glacial deposits are sorted and stratified materials deposited by meltwater and include:

- outwash plains
- kettle holes
- eskers
- kames

 ONLINE

Head to www.brightredbooks.net to see the answers for the practice question below.

THINGS TO DO AND THINK ABOUT

(a) Fully describe the evidence that suggests the type of upland landscape shown in the photograph. **(3 marks)**
Hint – your answer should mention the features you can see in the photograph.

(b) Choose one feature of glacial erosion shown in the photograph and, with the aid of diagrams, explain how it was formed. **(3 marks)**
Hint: Your answer to (b) should mention glacial processes.

COASTAL EROSION

The British Isles have a very long coastline, along which there is a wide variety of coastal landscapes. Many of these areas attract thousands of visitors every year and we will look at the effects of these visitors in a separate chapter.

WAVES

Waves play a critical part in shaping our coastline because they erode, transport and deposit material.

Friction caused by wind on the surface of water transfers energy into each wave. Wave height is affected by:
- the **strength** of the wind
- the length of time the wind blows, i.e. the **duration** of the wind
- the distance of the sea that the wave crosses – this is known as the **fetch**.

As waves approach a coastline, more friction is created with the sea bed. This slows the base of the waves down, while the top of the wave moves faster, causing it to increase in height and then crash onto the beach.

There are two types of wave:
- **constructive** waves create gently sloping beaches as they push material onto the beach – this action is known as **swash**
- **destructive** waves create steeply sloping beaches as they drag material back from the beach –this action is known as **backwash**

PROCESSES OF COASTAL EROSION

There are four main processes of coastal erosion:
1. **Hydraulic action** – this is caused by the sheer force of waves; air is trapped in cracks and is compressed, creating pressure on rock faces and causing them to fracture and break over time.
2. **Abrasion/corrasion** – this occurs when waves throw stones, sand and other materials onto a cliff face.
3. **Attrition** – as rocks and pebbles from a beach bump into one another, they are broken down into smaller particles.
4. **Corrosion/chemical weathering** – certain types of rocks are greatly affected by the chemicals found in seawater.

Freeze–thaw action can also affect coastal areas above the high tide mark.

ONLINE

Learn more about coastal erosion by following the link at www.brightredbooks.net

CLIFFS AND WAVE-CUT PLATFORMS

Photo of the formation of a wave-cut platform at the base of a cliff.

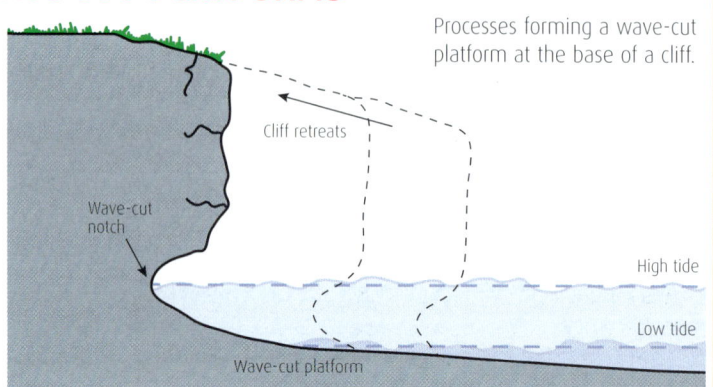

Processes forming a wave-cut platform at the base of a cliff.

Waves attack the base of cliffs between low and high tides. When waves break against the base of a cliff, hydraulic action and abrasion cause a wave-cut notch to develop. As the wave-cut notch becomes larger, the cliff above becomes increasingly unstable and, in time, the cliff will collapse. This process is then repeated, causing the cliff to become steeper and higher, with a large wave-cut platform developing at its base.

HEADLANDS AND BAYS: DISCORDANT COASTS

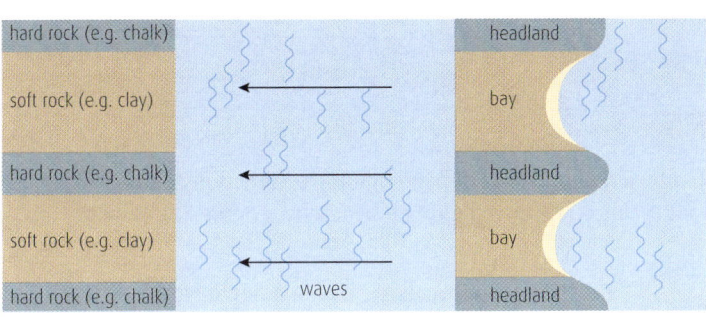

Formation of bays and headlands from alternating layers of hard and soft rocks.

Headlands and bays, Shetland.

Headlands and bays form along coasts that have alternate bands of harder (resistant) and softer (less resistant) rocks. Resistant rocks such as chalk take longer to erode, leaving a headland that juts out into the sea, whereas softer rocks such as clays are eroded more quickly, forming a bay. The headland will then become more vulnerable to hydraulic action and abrasion, and will be eroded faster.

COVES: CONCORDANT COASTS

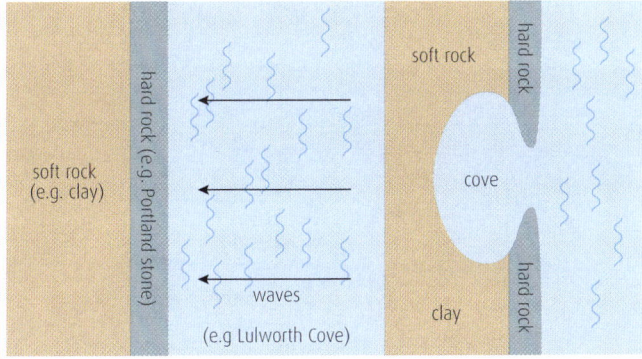

Formation of a cove from bands of hard and soft rocks parallel to the sea.

Lulworth Cove, south Dorset coast.

Coves such as Lulworth Cove in Dorset form when bands of rock of different hardness lie parallel to the sea. Here, a band of resistant rock (Portland limestone) lies close to the sea and a band of less resistant clay is found inland. Waves seek out faults in the hard rock and, by abrasion, corrasion and hydraulic action, eventually break through to the softer rock behind. These processes then erode the softer rock faster, leaving a circular cove with a narrow entrance where the sea enters. Waves are refracted within the cove and spread out to erode in all directions, helping to form this distinctive, almost circular shape.

The Old Man of Hoy in the Orkney Islands.

CAVES, ARCHES AND STACKS

This set of features shows the progressive erosion of a headland and cliff. Headlands contain areas of weakness and these are the first to be attacked by the sea via hydraulic action (**1**). Over time, this weakness is enlarged to form a sea cave (**2**). Hydraulic action and abrasion continue to attack the sea cave, causing it to break through the back of the headland to form an arch (**3**). Continued erosion of the arch by wave processes, chemical weathering and the effects of freeze–thaw cause the arch roof (**4**) to become unstable and it collapses, leaving a stack (**5**). Further undercutting of the stack causes it to collapse leaving a stump, usually only visible at low tide (**6**).

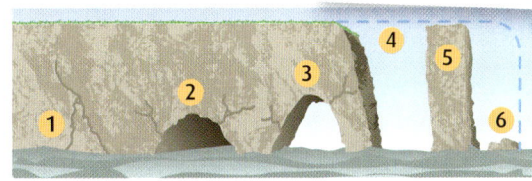

Formation of caves, stacks and arches.

 THINGS TO DO AND THINK ABOUT

These coastal features are all found in the British Isles. Think about a coastal location you have visited. Can you identify any of the features mentioned here in that landscape?

✔ **ONLINE TEST**

Test yourself on coastal erosion at www.brightredbooks.net

COASTAL DEPOSITION

LONGSHORE DRIFT

Waves usually approach a beach at an angle depending on the direction of the prevailing wind. This means that material is transported **along** a beach by a process known as **longshore drift**:

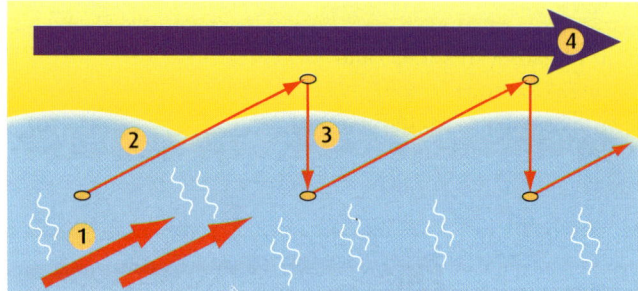

Longshore drift along a beach.

1 waves approach the beach at an angle as a result of the prevailing wind

2 sand particles/pebbles are moved up the beach at this angle – **swash**

3 the sand/pebbles then fall straight back down the beach – **backwash**

4 the sand moves laterally along the beach – **longshore drift**

Spits

If a coastline changes direction, the beach will continue to grow by longshore drift to form a **spit**. This feature grows in stages and curves upwards when the wind changes direction.

Spurn Head spit, Yorkshire.

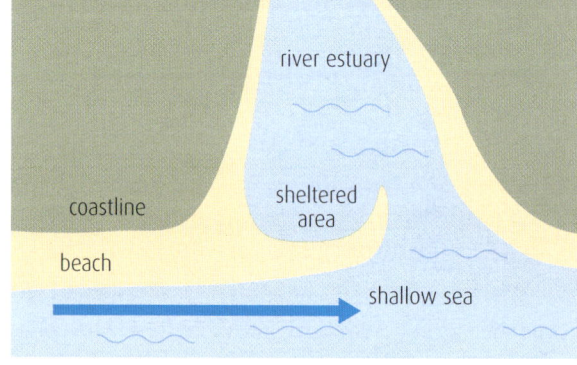

Formation of a spit.

Bars and tombolos

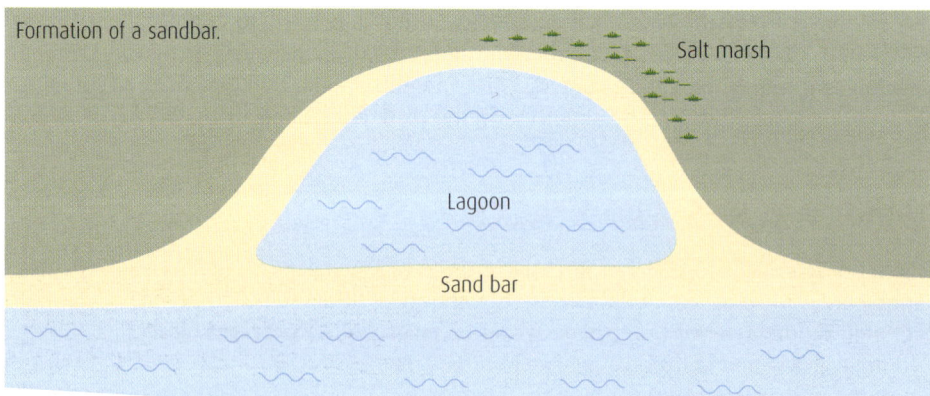

Formation of a sandbar.

Example of a sand bar: Slapton Sands, Devon.

A **bar** is formed when a barrier of sand stretches across a sheltered bay. This will only happen if there is no large river to generate water movement. The area of water is known as a **lagoon** and will host a wide variety of plants and wildlife. Over time, the water behind the bar may eventually dry up.

Sometimes a spit becomes detached from the mainland so will form an offshore bar or barrier islands. An example of this is the area of water behind Chesil Beach in Dorset. The photograph on page 17 shows how longshore drift has extended the beach to join up with the Isle of Portland in the distance. This feature is known as a **tombolo**. Spits can also do this.

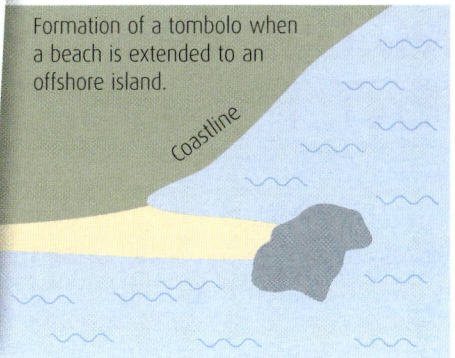

Formation of a tombolo when a beach is extended to an offshore island.

Coastline

Chesil beach sand bar joining up with the island of Portland, Weymouth.

COASTAL FEATURES ON ORDNANCE SURVEY MAPS

Coastal features can be identified on Ordnance Survey maps.

Marshes

Longshore Drift

Lagoon

Beaches are evidence of deposition

Stacks and stumps

Different rock strengths form headlands and bays

Cliff faces

A 'Down' shows this is chalk landscape in southeast England

Presence of groynes may show problems with erosion

Coastlines of erosion and deposition on a map.

- Features of both erosion and deposition are often shown on the same map extract.
- Erosional features include wave-cut notches and caves.
- Depositional features include spits, tombolos, bars and lagoons.
- Cliffs may be identifiable by very close contour patterns.
- Coves may indicate alternating bands of weak and resistant rock running parallel to the coast.

ONLINE

Head to www.brightredbooks.net to learn more about features of coastal deposition.

DON'T FORGET

Remember to learn named examples of features for the exam!

ONLINE

Follow the links at www.brightredbooks.net to revise coastal erosion and Ordnance Survey maps.

DON'T FORGET

Your teacher can provide you with a variety of Ordnance Survey maps to practise identifying both erosional and depositional coastal features.

ONLINE TEST

Want to test yourself on coastal deposition? Head to www.brightredbooks.net

ONLINE

Head to www.brightredbooks.net to see a model answer.

THINGS TO DO AND THINK ABOUT

You will have to explain each process for the exam when answering a question about feature formation. Lists of processes will only gain a maximum of two marks. This is your chance to 'show off' to the examiner what you know.

1 With the aid of annotated diagrams, fully **describe** and **explain** the various stages and processes involved in the formation of **either** a stack **or** a sand bar.

RURAL LAND USE CONFLICTS AND THEIR MANAGEMENT 1

The landscape in upland rural and coastal areas is directly affected by **human activity**. These areas present a variety of physical, economic and social opportunities for a range of land users and many challenges and **conflicts** arise between these different groups.

TOURISM

The main source of conflict in rural areas is between **tourists** and **other land users**. Tourists bring both benefits and problems to these areas. Strategies are implemented to minimise the impact of land use conflicts, with variable effectiveness.

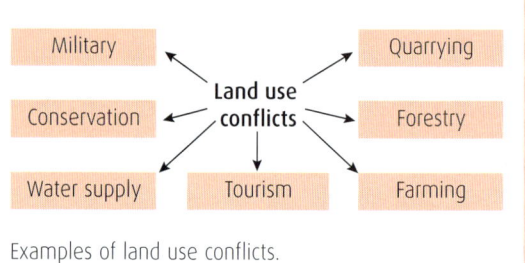

Examples of land use conflicts.

SOCIAL AND ECONOMIC OPPORTUNITIES IN UPLAND RURAL LANDSCAPES

Social opportunities	Economic opportunities
Mountaineering, hill walking	Hill farming (sheep)
Forest walks, picnicking, orienteering courses	Development of hotels, campsites, bunkhouses
Water sports, sailing	Hydro-electric power, supply of water
Fishing	Quarrying
Nature conservation	Forestry plantations
Tourism and associated activities	Tourism and associated employment
	Energy generation, e.g. wind turbines

In the UK, upland areas such as the National Parks provide a range of social and economic opportunities for people. The opportunities provided will vary depending on the nature of the area and the physical landscape.

BENEFITS AND PROBLEMS CREATED BY TOURISM IN RURAL AREAS

Benefits	Problems
Tourists spend money in the local area boosting the local economy	Day-trippers may spend little money in the local area
The multiplier effect spreads through the local economy	
Local businesses, such as restaurants, cafes and hotels, thrive in honey-pot sites	Seasonal congestion can cause parking problems for local residents
	Local convenience businesses may decline as tourist-related shops thrive
Tourism creates job opportunities, including long-term and seasonal work	Seasonal tourism and jobs can leave tourist areas as ghost towns in the off-season
	Seasonal work may attract migrants and students rather than providing jobs for local residents, affecting the long-term economy
Local infrastructure may benefit from tourism as bus timetables and roads are adapted to suit the large influx of visitors	Increased tourist traffic may interfere with traffic flow, increasing traffic accidents
Second homes are often purchased by visitors, boosting the local economy	Local people struggle to own their own home as second homes push up property prices
	Local services decline because tourists do not use local shops or primary schools

CASE STUDY: THE ENGLISH LAKE DISTRICT

The Lake District is England's largest National Park. It is easily accessible by motorway, which makes it attractive to tourists who live within a couple of hours drive away in cities such as Manchester and Liverpool.

Farming

Farming, particularly sheep farming, has always been a major industry in the area. It is difficult to farm here as a result of the cool, wet climate and thin soils, but sheep-farming is important economically and is essential in preserving the natural landscape. Farms are extensive and require subsidies to survive. Many farmers have diversified into tourism to supplement their income.

Mining/quarrying

Slate mining is a traditional industry, but it too has diversified, for example, the development of the slate mine at Honister as a tourist destination. Many products are made from the traditional green slate and there are also tours and high-level walks.

Recreation

The main activity in the Lake District is walking, along with a wide variety of active and passive pursuits. Water activities are popular on the many lakes.

Map of the English Lake District in Cumbria.

ENVIRONMENTAL ISSUES AND CONFLICTS

Quarrying

Tourists dislike the eyesore of quarries. Lorries are noisy and cause congestion on narrow roads. Dust from quarries causes visual pollution and it is difficult to obtain planning permission for new quarries from the National Park Authority. Water supplies may become polluted.

Tourism

Tourism is beneficial economically, but residents dislike the congestion and parking is almost impossible, particularly on Bank Holiday weekends. Footpaths have been eroded, leaving visual scars on the hillsides. Litter detracts from the appearance of an area and can harm both livestock and wildlife. Tourists sometimes wander over cultivated land, leaving gates open, worrying sheep and damaging walls. House prices have increased as a result of both incomers and the demand for second homes. Young people from the local area are often forced to move away because of increased house prices.

Reducing the impact of tourism

The National Park Authority has attempted to ease these tensions by, perhaps surprisingly, **removing** litter bins, so that visitors have to take their litter home. In remote areas it is difficult to empty bins regularly and they quickly become eyesores with high visitor numbers.

Fix the Fells, a partnership project involving the National Park Authority, the National Trust and Natural England, has been addressing the problems of tourism in recent years.

The National Park Authority promotes the use of public transport and has a GoLakes Travel Programme. It has worked with the Department of Transport, Cumbria County Council and Cumbria Tourism on a wide range of transport activities, including enhanced public transport services, cycle hire and improvements to routes, better information about travel options, marketing and PR activities, and training for the tourism industry.

THINGS TO DO AND THINK ABOUT

The website www.lakedistrict.gov.uk has a wealth of information for visitors on the area's attractions, activities and conservation work.

VIDEO LINK

Learn more about Fix the Fells by watching the clip at www.brightredbooks.net

DON'T FORGET

In the exam, you are expected to give named examples of organisations that support the management of tourism/farming in the environment.

ONLINE TEST

Want to revise your knowledge of this topic? Take the test at www.brightredbooks.net

RURAL LAND USE CONFLICTS AND THEIR MANAGEMENT 2

DORSET COAST

Tourism

The stunning scenery of the Jurassic Coast has led to it becoming Britain's first World Heritage Site.

Map of the Jurassic Coast, Dorset.

ONLINE

Learn more about this area by following the link at www.brightredbooks.net

The Jurassic Coast has 155 km of unspoilt cliffs and beaches, which are easily accessible through gateway towns (see above map) and the South-West Coastal Path.

Farming

Agriculture is the major land use – around three-quarters of the land is used for farming. A wide variety of crops is produced, along with pastoral farming. However, this has recently become less profitable and only supports around 3% of employment in the area.

Military

There has been a military presence in Dorset for a long time. This provides both employment for the local population and training facilities.

Recreation

Tourism is the most important economic activity. The ports of Poole, Portland and Weymouth generate international trade and tourism. Approximately 40 000 locals are employed in tourist-related jobs, and visitors are evenly spread between day-trippers and longer-stay guests.

The 2012 Olympic Games had a positive effect on the area, raising its profile as a result of the sailing events held there. The area experienced increased investment in infrastructure and a growth in the marine leisure sector. This should continue to have a positive effect on local businesses and tourism.

ENVIRONMENTAL ISSUES AND CONFLICTS

Similar to the Lake District, there are also many negative effects of the economic/social use of this area. The greatest conflicts are around tourism.

Tourism: positive effects

Lulworth is a picturesque village on the South-West Coastal Path:

- About 750 000 people visit Lulworth each year.
- 35% of visitors come in a six-week period during July and August.

contd

- Only 10% of visitors come during the winter months of November to February.
- 95% of Lulworth's visitors are day-trippers.
- Most (over 90%) come by car or coach.
- The Heritage Centre is Dorset's second most visited tourist attraction, and its most visited free attraction.
- The footpath between Lulworth and Durdle Door is the busiest one-mile stretch of the whole South-West Coastal Path.
- It is surrounded by outstanding scenery and wildlife habitats, for example, the Cove, Stair Hole, cliff path views, Durdle Door.
- The tourist infrastructure includes cafes, hotels, B&Bs, ice-cream kiosks, a heritage centre, various shops, a holiday park and a youth hostel.
- The village has thatched cottages.
- The nearby Fossil Forest is an important geological Site of Special Scientific Interest.

Tourism: negative effects

- The most obvious impact on local people is the volume of traffic, resulting in congestion and illegal parking – noise and litter are major complaints and there are many second homes.
- Military activities such as firing ranges cause areas of the coastal path to be closed and tourists complain about noise pollution and restricted access when manoeuvres are taking place – weapons firing has caused landslides in some areas.
- Poole Harbour experiences high volumes of use (ferry traffic, water activities and an oil terminal) – the many uses are zoned so that users are safe and the fragile marine environment can be maintained.

Tourism: impacts and management

Although there are clear economic benefits to tourism, there have been compromises within these areas.

Military compromises include:

- Access is permitted to the ranges at weekends and in busy holiday periods.
- Roads are kept open during the busiest holiday periods.
- Noise levels associated with firing are reduced at these times.

Around Lulworth:

- The Lulworth Estate manages a car park that accommodates over 500 vehicles.
- A mini-roundabout provides easy access to the car park.
- The estate subsidises a bus service from the local railway station to encourage visitors to leave their car at home.
- Litter bins are not provided – visitors are encouraged to take litter away.
- Climbing on the cliffs is banned to prevent damage to the fragile chalk and limit disturbance to wildlife – in some areas where footpath erosion is particularly bad, parts of the South-West Coast Path have been redirected.

DON'T FORGET

When revising, learn **named** examples.

THINGS TO DO AND THINK ABOUT

Try summarising the impacts for the two areas studied in this chapter (Lake District and Dorset coast) into a table.

ONLINE TEST

Test your knowledge of this topic at www.brightredbooks.net

COASTAL DEFENCES AND MASS MOVEMENTS

Concrete walls at Brighton Marina.

COASTAL DEFENCES

Coastal erosion is a major problem on the Dorset coast. Many of the towns on the actual coast rely on their beaches to attract visitors and have had to put in a number of coastal protection systems. These are divided into 'hard' and 'soft' defences.

Hard defences

Hard defences include concrete walls, stone walls/rip-rap/rock armour and groynes.

Unfortunately, hard defences, while limiting erosion in one area, often increase the rate of erosion elsewhere. The cost of constructing and maintaining sea walls can be very high.

Soft defences

Soft defences include beach nourishment and stabilisation, which minimise human interference.

Beach nourishment means bringing in sand to add volume to the beach. By maintaining the amount of material present, the beach remains wide and erosion occurs more slowly.

Vegetation can be planted on dunes to help to reduce erosion. Dawlish Warren sand spit in Devon is protected in a number of ways, including dune stabilisation.

Stone walls/rip-rap/rock armour near Bournemouth.

Groynes to limit longshore drift in Swanage Bay.

Protection of sand spit at Dawlish Warren by planting marram grass.

ONLINE

For further reading about coastal defences, follow the links at www.brightredbooks.net

MASS MOVEMENT

The processes that move large amounts of soil, stones and rock (known as **regolith**) downhill under the force of gravity are known as mass movement. They are influenced by:

- the type of rock and its structure – whether it is porous or impermeable, weak or strong, jointed or sloping
- how much water is present and whether the debris is saturated
- the amount of surface vegetation
- the angle of slope – steep slopes encourage faster movement.

contd

Mass movement is important in landscape formation and often works with other erosion and transport processes. The removal of material produced by a rockfall, for instance, may be carried out by a river. Types of mass movement include:

- **landslides** – rapid movements of regolith on slopes when the underlying rock can no longer support the weight above – they occur where saturated bedding planes, lubricated by heavy rain, lie between contrasting layers of rock; **slumps** occur in weaker rocks and involve **rotational movement**

- **rockfalls** – very rapid movements on steep slopes or cliffs when freeze–thaw or exfoliation processes are at work – debris will accumulate at the foot of the cliff, as a **scree** or **talus** slope.

Examples of mass movement on slopes: rockfalls and slumps.

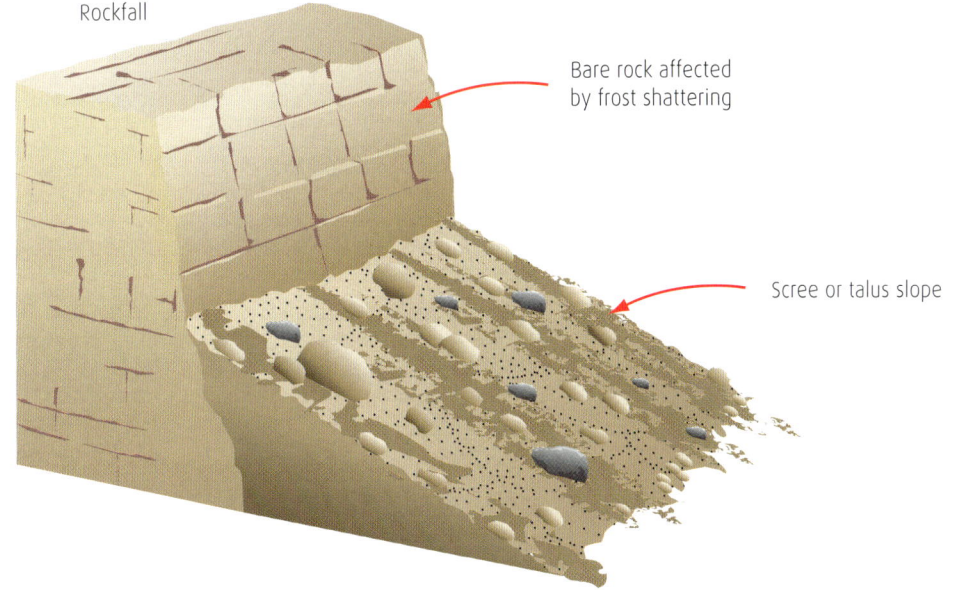

Rockfall

Bare rock affected by frost shattering

Scree or talus slope

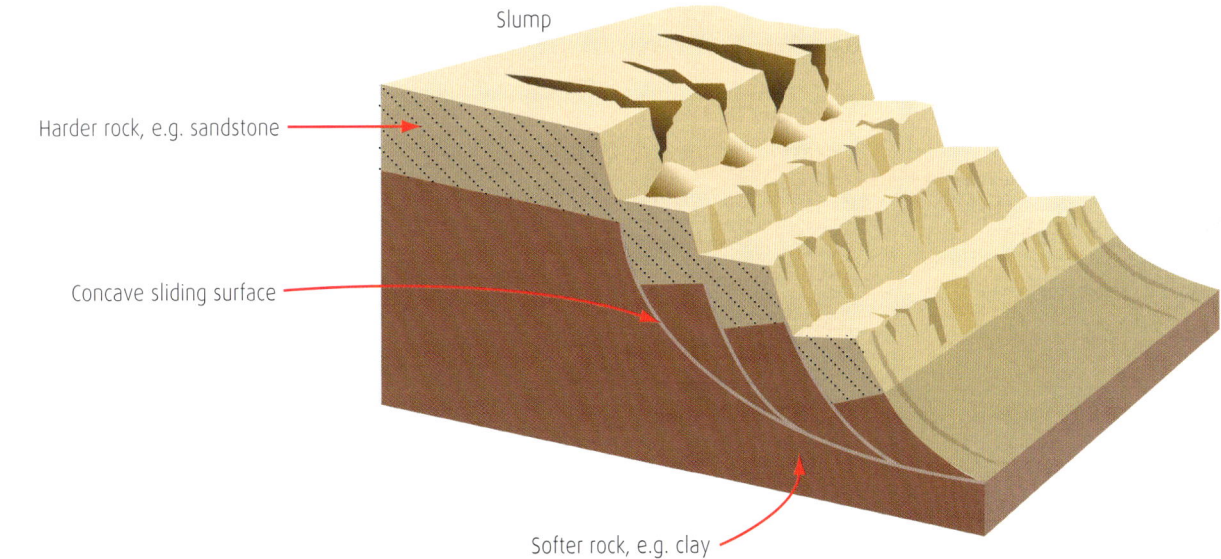

Slump

Harder rock, e.g. sandstone

Concave sliding surface

Softer rock, e.g. clay

 **DON'T FORGET**

Questions solely about mass movements are rare (although not unknown), but knowing how they are caused may prove useful in answering other questions.

 ONLINE

Details of the disaster caused by mass movement at Aberfan in Wales can be found at www.brightredbooks.net

 THINGS TO DO AND THINK ABOUT

You will have to explain each process for the exam when answering a question about feature formation. Lists of processes will only gain a maximum of two marks. This is your chance to 'show off' to the examiner what you know.

 ONLINE TEST

Test yourself on this topic online at www.brightredbooks.net

CASE STUDY AREAS

SUMMARISING INFORMATION FOR REVISION PURPOSES

A good way to summarise your information is in a table.

Example

	Lake District: an upland, glaciated area		Dorset: a coastal area	
Social and economic opportunities	**Social** Mountaineering, hill walking, water sports, mountain biking, picnics	**Economic** Quarrying, e.g. slate, sheep-farming, hydro-electric power, tourism	**Social** Mountain biking, photography, picnics, walking, water sports	**Economic** Farming (pastoral), tourism, oil refinery and petrochemical plant
Limits to economic opportunities	Poor climate – cool, high rainfall and few sunlight hours affects farming Steep slopes make cultivation difficult		No motorways	
Impact of tourism	**Positive impacts of tourism** Money brought into the local economy, multiplier effect, pubs and restaurants, variety of accommodation, jobs, infrastructure improvements, run-down properties/buildings improved as second homes, close to large centres of population, i.e. visitors nearby **Negative impacts of tourism** Mainly day trips; honey-pots (Windermere), limited types of shops (to suit tourists, not locals), seasonal employment, congestion, parking issues, second homes, gates left open, litter, footpath erosion		**Positive impacts of tourism** Money brought into local economy, multiplier effect, pubs and restaurants, variety of accommodation types, jobs, infrastructure improvements, run-down properties/buildings improved as second homes, 2012 Olympic Games **Negative impacts of tourism** Mainly day trips; honey-pots (Lulworth Cove), limited types of shops (to suit tourists, not locals), seasonal employment, congestion, parking issues, second homes, gates left open, litter, footpath erosion, marine pollution in Poole Harbour	
Management of protected or sensitive areas	National Park status helps to manage/control impact of economic activities and preserve landscape National Trust manages some visitor areas, e.g. Beatrix Potter Cottage Forestry Commission sites, e.g. Grizedale Forest Zoning of Lake Windermere for different activities Ethos of co-operation and compromise		Area has Heritage Coast status and there are designated National Nature Reserves and Sites of Special Scientific Interest National Trust is involved, e.g. Corfe Castle Local voluntary organisations repair footpaths Ethos of co-operation and compromise Conflicts of military use with tourism, but compromises on firing times	

ANNOTATING DIAGRAMS

The exam may ask you to annotate diagrams of features in coastal or glacial areas.

 **ACTIVITY**

Complete the key below the field sketch (Diagram A) to identify the main erosional features of the coastal landscape indicated by the letters A to F.

Diagram A: sketch of selected coastal erosion features.

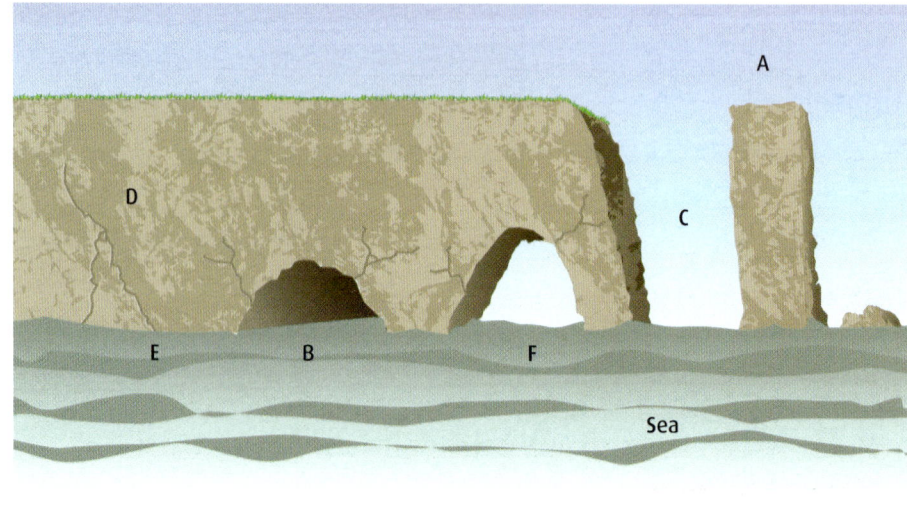

Key   A _____   B _____   C _____

　　　　D _____   E _____   F _____

THINGS TO DO AND THINK ABOUT

Your teacher may have issued you with different case studies. It is good practice to summarise your own notes from class as your teacher will have gone into these in great depth. Choose one of the following summarising techniques:

- spider diagrams or mind-maps
- bullet points
- tables
- summary cards

Whichever technique you choose, you can include a lot of information for revision. Cards will not allow huge amounts of information, so one card per sub heading may work. If you are using a sheet of A4 paper, your table could look like this:

Case study	Dorset Coast
Key features: coastal erosion	Cave: Tilly Whim Caves, near Swanage Arch: Durdle Door Stack: Old Harry Stump: Old Harry's Wife Bay Headland
Diagrams	Caves, arches and stacks diagram (annotate the diagram)
Key visitor areas	Lulworth Cove: a honey-pot Durdle Door: path from Lulworth to Durdle Door heavily eroded
Impact of tourism: positive effects	Employment, accommodation
Impact of tourism: negative effects	Seasonal employment, parking problems, heavy congestion
Other human impacts	Oil, military use, variety of activities in Poole Harbour
Management of impacts	Designations, e.g. National Nature Reserves and Sites of Special Scientific Interest, World Heritage Site

Now do the same for your glaciated area.

Practice question:

1 For a coastal area you have studied:

   (a) describe the management strategies used to manage conflicts

   (b) comment on the effectiveness of these strategies.

For the exam, you need to know the following for the Lithosphere topic:
- Formation of erosional and depositional features in glaciated and coastal landscapes.
- Rural land use conflicts and their management related to glaciated and coastal landscapes.

ONLINE

The Jurassic Coast link at www.brightredbooks.net will lead you to an excellent website, detailing the attractions of this beautiful coastline and management strategies.

DON'T FORGET

Give named examples of tourist villages or towns and tourist facilities for your two case study areas.

ONLINE

Head to www.brightredbooks.net to see the answers for this question.

ONLINE TEST

Head to www.brightredbooks.net for tests on this chapter.

HYDROLOGICAL SYSTEMS

GLOBAL HYDROLOGICAL CYCLE

The global hydrological cycle is powered by solar energy. It is a closed system because it has a fixed amount of water.

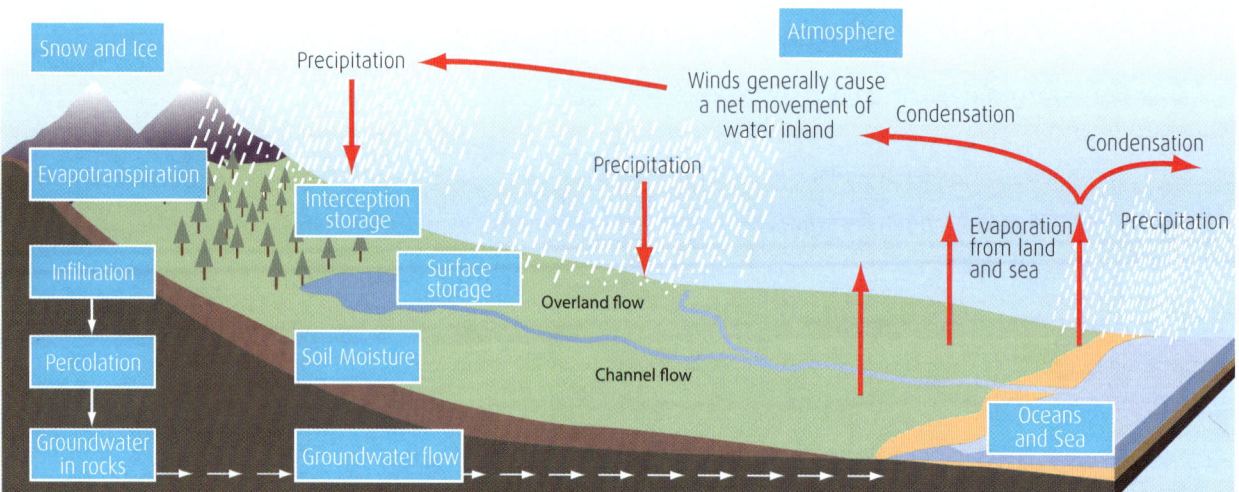

The global hydrological cycle.

Most of the water (99%) is stored in the seas, ice caps and glaciers and only about 1% is in circulation at any one time. You should be able to describe the system shown in the diagram.

Description of the global hydrological cycle

The Sun causes **evaporation** from oceans, lakes, rivers and other areas of surface water. It also causes **transpiration** (the loss of water) from plants. As this moisture rises, it cools to form clouds, a process known as **condensation**. When the air cools further, **precipitation** forms and falls to the ground, contributing to **surface run-off**, which makes its way back to the oceans. Some of the moisture enters the soil via **infiltration**, moving through the soil as **through-flow** and **percolating** into the **groundwater** store.

The main **input** into this system is precipitation. The type of precipitation (rain, hail, snow, sleet) and its intensity, duration and frequency will affect the movement of water through the system.

Water is stored in lakes, channels, underground and in rocks. The amount of water that can be stored depends on the soil porosity and the permeability of the rock. Water can also be intercepted and stored in plants.

Water is **transferred** through the system in a variety of ways: run-off can be either **stream flow** or **overland flow**. Stream flow occurs in river/stream channels, whereas overland flow may be as **sheet wash** or **rills**. Water that moves through the lower soil levels (throughflow) will make its way back into rivers. **Groundwater** movement is very slow. Plants have **stem flow** to allow movement back into the system.

Output from the system takes two forms: the main output is from rivers into the sea; **evapotranspiration** plays a lesser part.

DON'T FORGET

There are many different processes operating in the hydrological cycle, but it is basically a more detailed water cycle than the one you learned in primary school – do not be put off by the more scientific terms.

FACTORS AFFECTING THE HYDROLOGICAL CYCLE

Physical factors

The amount of water in a drainage basin varies according to a number of **physical** factors:

- type of precipitation
- intensity of precipitation
- duration of precipitation
- frequency of precipitation
- number of streams within the drainage basin
- soil type – sandy soils are more porous and allow water to pass through more easily
- vegetation cover – trees slow down interception rates more than grass and transpiration rates also vary
- rock type – chalks have a higher infiltration rate than clays
- temperature – evaporation rates are higher in warmer areas

Human factors

There are also a number of **human factors** that affect the hydrological cycle:

- **Deforestation** – cutting down trees increases run-off, decreases evapotranspiration (and therefore cloud formation) and leads to more extreme river flows as water is no longer intercepted and stored by trees.

- **Irrigation** – taking water from a river or underground store can reduce river flow, lower water tables and increase evaporation/evapotranspiration by placing water in surface stores (ditches/canals), or by crops removing water from the cycle as they grow.

- **Urbanisation** – removal of natural vegetation and replacement with impermeable surfaces and drains speeds up overland flow and evaporation and can lead to higher river levels; it also decreases the amount of water returning to groundwater storage, possibly reducing the water table.

- **Mining** – silting-up of lakes, rivers and reservoirs leads to reduced storage capacity in these areas; mining may also lead to reduced vegetation cover and increased run-off, higher evapotranspiration and cloud formation, thus altering the rainfall patterns.

- **Dam building and reservoirs** – these increase the amount of water available for evaporation and affect the local climate.

- **Afforestation** – this increases the interception rates in the drainage basin.

- **Vegetation/no cover** – interception and infiltration rates are affected if there are crops in a field rather than bare soil after harvesting.

THINGS TO DO AND THINK ABOUT

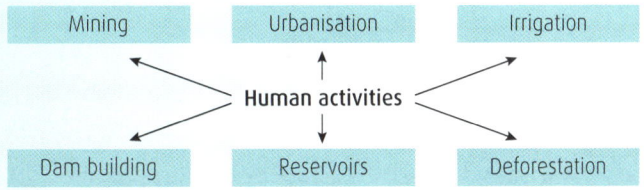

(a) Describe how human activities, such as those in the diagram, can affect the hydrological cycle.

(b) Explain these effects.

STORM HYDROGRAPHS

STORM HYDROGRAPHS: AN OVERVIEW

To compare rivers, we need to look at their **discharge**, that is, the amount of water in the river passing a certain point every second. Discharge is measured in **cumecs** (cubic metres per second) and is calculated by multiplying the speed of the river by its cross-sectional area.

At various points in a drainage basin, the discharge will be measured at a **gauging station**. It can then be compared with the precipitation by drawing a **storm** (or **flood**) **hydrograph**.

A storm hydrograph contains two graphs: a **bar graph** showing the rainfall and a **line graph** showing the river's discharge before, during and after a storm.

Example of a storm hydrograph.

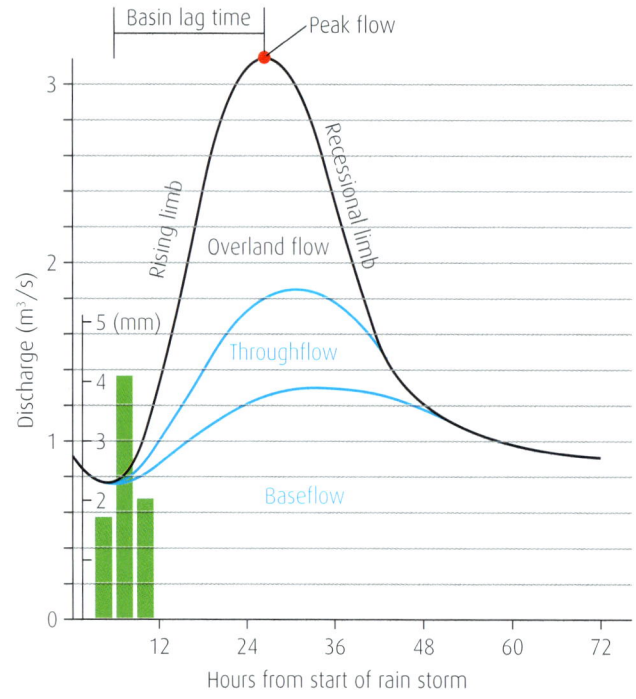

DON'T FORGET

You have to be able to annotate the various parts of a hydrograph and to analyse what they tell you.

DON'T FORGET

Rainfall on the hydrograph is shown as a bar chart and the discharge is shown as a line graph. There are two separate scales on the y-axis for the rainfall and discharge, so be sure to distinguish between them.

FEATURES OF HYDROGRAPHS

The features shown on a hydrograph include the following:

- **Precipitation** – normally related to a specific event such as a storm and shown against a **time-scale** on the *x*-axis.
- The normal level of the river, known as the **baseflow**.
- The **discharge** of the river plotted against time.
- The **peak flow** discharge for the storm period.
- The gap between the peak rainfall and peak flow, known as the **basin lag time**; this reflects the direct channel precipitation plus the overland flow and throughflow, which take longer to reach the river.
- The **rising limb**, which shows how quickly the water reaches the river channel.
- The **recessional** or **falling limb**, which shows the decrease from peak flow.
- The **storm flow** – the additional discharge produced by the storm event.
- The **bankfull discharge level** above which the river will be in flood.

VARIATION BETWEEN DRAINAGE BASINS

Storm hydrographs vary between drainage basins. These differences may include the following:

- The **size** of the drainage basin – a larger basin receives more precipitation and will therefore have a larger run-off. The lag time will also be longer because the distance the water has to travel to reach the channel will be larger and there may also be more channels within the drainage basin. Smaller basins have shorter lag times because the water takes less time to reach the river.

- **Shape** – circular basins drain more quickly; water takes longer to reach the river from a long, narrow basin.

- **Type of precipitation** – in heavy storms, rainfall is often far greater than the infiltration capacity of the soil, leading to overland flow and rapid rises in river levels. During long periods of **rainfall**, the ground becomes saturated and overland flow increases. If there is **snowfall**, the potential discharge of a river is held in storage until the snow melts – rapid melting can lead to flooding.

- At high **temperatures**, rates of evapotranspiration are high and there are reduced amounts of discharge. During periods of low temperature, water is stored as ice and snow and river discharge is reduced.

- **Vegetation** reduces discharge because it intercepts precipitation and adds to evapotranspiration. The plant roots take up water, reducing throughflow. Interception is less in winter in the UK because deciduous trees shed their leaves. However, coniferous forests have fairly stable discharge rates. Flooding is more likely in deforested areas.

- Drainage basins with a higher **drainage density** have a greater chance of flooding.

- **Steepness** of slopes – steeper slopes (high **relief**) give a steeper rising limb.

- **Rock/soil type** – impermeable rocks produce greater overland flow as infiltration rates are lower. The amount of soil moisture storage and the rate of throughflow affect infiltration. Larger pores in sandy soils allow greater drainage and limit flood risks.

- **Urban** areas – infiltration rates are low due to tarmac surfaces and the lag time is lower. The rising limb is steep as water reaches the river channel more quickly via gutters and urban drainage systems.

 DON'T FORGET

Draw a hydrograph and add its key features. This will help you in the exam. If you know everything on the diagram, then this will encourage you to write down enough points for the examiner.

ONLINE

The link at www.brightredbooks.net has information on storm hydrographs for you to revise.

THINGS TO DO AND THINK ABOUT

Storm hydrographs are important tools for town and country planners throughout the world. They can be used to predict river levels and are important in flood control and in areas where droughts are common.

Explain the differences in discharge between urban and rural hydrographs following a heavy rain storm.

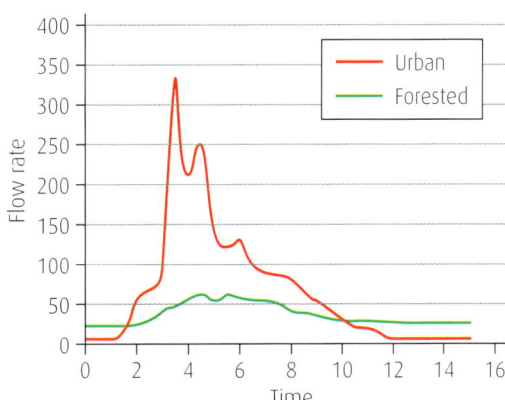

Comparison of storm hydrographs for urban and forested areas.

 ONLINE TEST

Test yourself on storm hydrographs at www.brightredbooks.net

ONLINE

Head to www.brightredbooks.net to see a model answer.

For the exam, you need to know the following for the Hydrosphere topic:
- hydrological cycle within a drainage basin
- interpretation of hydrographs

GLOBAL HEAT BUDGET

The Sun is the Earth's source of energy. It 'powers' the atmosphere and makes life on Earth possible.

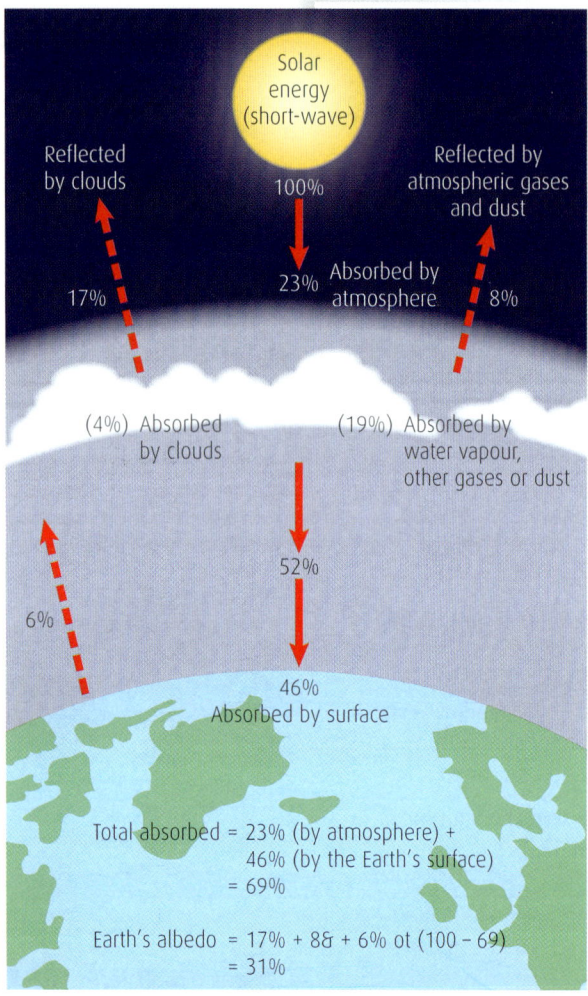

Total absorbed = 23% (by atmosphere) +
                 46% (by the Earth's surface)
               = 69%

Earth's albedo = 17% + 8& + 6% ot (100 – 69)
               = 31%

GLOBAL HEAT BUDGET: AN OVERVIEW

The incoming energy (**short-wave energy** or **insolation**) is balanced by the outgoing energy (**long-wave** or **infra-red radiation**). The balance between the input and output is the **global heat budget**.

Energy from the Sun passes through the atmosphere to the Earth's surface. Some radiation is reflected by clouds (17%), some is scattered by gas particles (8%) and some is reflected by the Earth's surface (6%). This is known as the **albedo**. A total of 31% of the incoming solar radiation is lost.

A total of 4% of the incoming radiation is absorbed by clouds and 19% is absorbed by dust, water vapour and gases.

This means that 54% of the incoming radiation is absorbed or reflected, so only 46% reaches the Earth's surface. About 6% is then reflected by the Earth's surface as long-wave radiation, so 40% is left to heat the Earth's atmosphere via gases and water vapour. This is the greenhouse effect, which is necessary for our survival – without it the Earth would be 30–40°C colder.

The Earth's global heat budget.

DON'T FORGET

In the exam, quote actual values for the global heat budget. Sometimes the values vary slightly. Don't let this throw you.

GLOBAL HEAT BUDGET

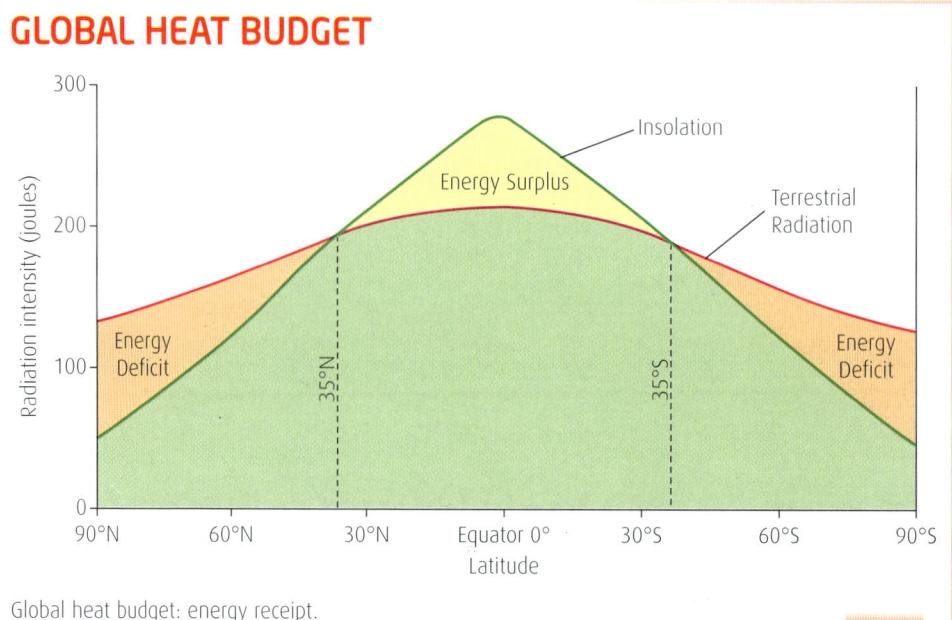

Global heat budget: energy receipt.

contd

ONLINE

Learn more about the BBC programme *Orbit* by following the link at www.brightredbooks.net

More energy is received at the Equator than at the poles. The reasons for this are as follows:

- The **shape** of the Earth. Energy from the Sun approaches the Earth in straight lines and hits the area between the Tropic of Cancer and the Tropic of Capricorn (A–B) at right angles; further north, nearer the North Pole (X–Y), the angle is greater. Therefore, the energy absorbed at the tropics is more intense or stronger, whereas at the poles it will be less intense or weaker.

- The **size** of the area being warmed. The energy absorbed at the tropics has a smaller area to warm than at the poles, so the tropics are warmer.

- The **effect of the atmosphere**. There is more atmosphere for energy to pass through at the poles, so more energy will be absorbed and reflected by clouds, dust, gases and water vapour at the poles than in the tropics.

- The **colour** of the surface. The land surface at the poles is mainly white, so more energy will be reflected than at the darker tropics, which absorb more energy – this can be seen on satellite images of the Earth from space.

- The Earth's **revolution around the Sun**. The Earth sits at an angle in space in relation to the Sun so, depending on the Earth's position as it revolves around the Sun, certain areas will receive more energy. The variation in insolation is most extreme at the poles, where there are approximately six months of light and six months of dark.

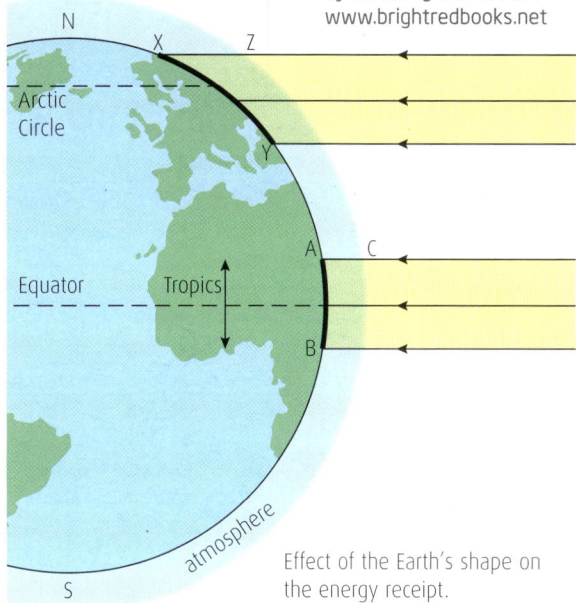

Effect of the Earth's shape on the energy receipt.

 DON'T FORGET

Practise drawing a globe to show areas of surplus and deficit heat and annotate it with the reasons for the differences.

GLOBAL INSOLATION

During the summer in the UK, the Sun is higher in the sky, so temperatures are higher and the days are longer. In winter, the Sun is lower in the sky, the intensity of its heat is less and the days are shorter.

Differences in global insolation explain why the tropics receive more energy than the poles.

Nature tries to even out these global differences to achieve a balance. Without the following processes, warm areas would become warmer and cold areas would become colder:

- **Atmospheric circulation** – wind, storms and hurricanes transfer around 80% of the total energy.

- **Oceanic circulation** – the remaining 20% is transferred by oceanic currents.

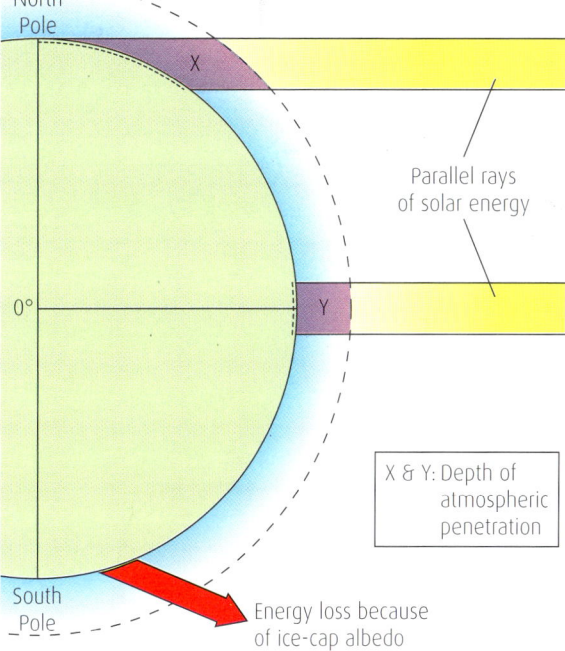

Parallel rays of solar energy

X & Y: Depth of atmospheric penetration

Incoming solar radiation and the effect of latitude.

Energy loss because of ice-cap albedo

 ONLINE

Head to www.brightredbooks.net to see a model answer.

THINGS TO DO AND THINK ABOUT

The BBC programme *Orbit* shows a year in the Earth's orbit around the Sun and explains this concept really well.

Answer the questions below using *Orbit* and the effect of latitude diagram.

(a) Describe the areas of surplus and deficit solar radiation on the Earth's surface.

(b) Explain fully the reasons for these differences.

 ONLINE TEST

Head to www.brightredbooks.net and test yourself on this topic.

ENERGY TRANSFER 1

ATMOSPHERIC CIRCULATION

In an ideal situation, air from the tropics (the area of surplus) moves to the poles (the area of deficit). Nature, however, is not so straightforward and there is a three-cell model.

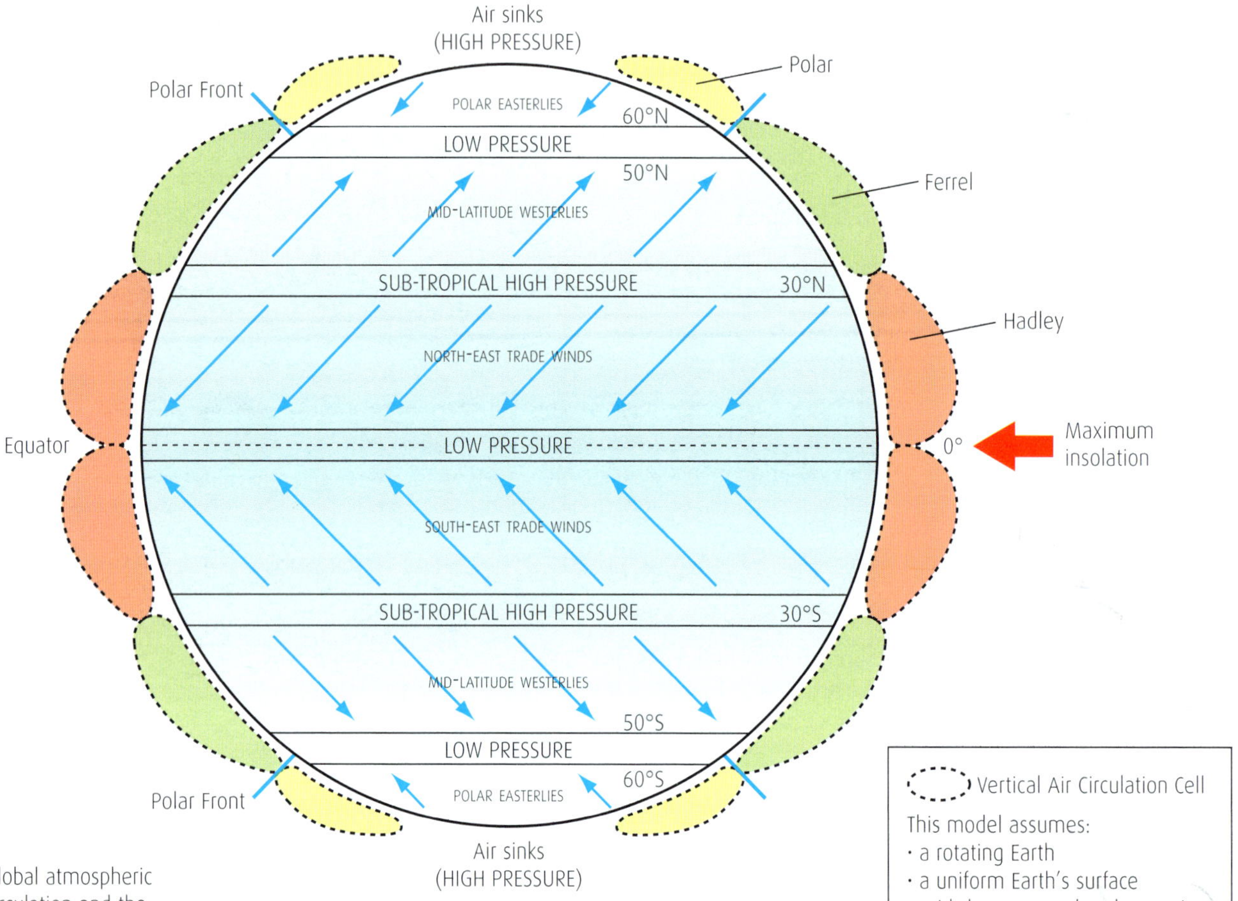

Air sinks
(HIGH PRESSURE)

Polar Front

Polar

POLAR EASTERLIES

60°N

LOW PRESSURE

50°N

Ferrel

MID-LATITUDE WESTERLIES

SUB-TROPICAL HIGH PRESSURE 30°N

Hadley

NORTH-EAST TRADE WINDS

Equator

LOW PRESSURE

0°

Maximum insolation

SOUTH-EAST TRADE WINDS

SUB-TROPICAL HIGH PRESSURE 30°S

MID-LATITUDE WESTERLIES

50°S

LOW PRESSURE

60°S

POLAR EASTERLIES

Polar Front

Air sinks
(HIGH PRESSURE)

Global atmospheric circulation and the three-cell model.

⟨ ⟩ Vertical Air Circulation Cell

This model assumes:
· a rotating Earth
· a uniform Earth's surface
· mid-day sun overhead at equinox

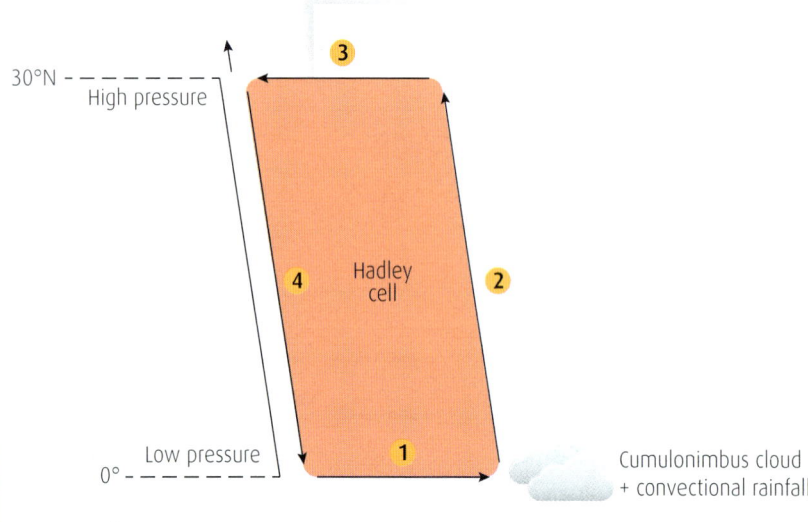

30°N
High pressure

3

Hadley cell

4   2

Low pressure
0°

1

Cumulonimbus cloud + convectional rainfall

A Hadley cell.

HADLEY CELLS

In the Hadley cells, warm, moist air rises quickly at the Equator, causing clouds to form and rain to fall. This is an area of **low pressure**.

The air then spreads north and south of the Equator before cooling and descending at 30°N and 30°S of the Equator. Where the air descends, an area of **high pressure** is created.

FERREL CELLS

The Ferrel cells lie between the Hadley and polar cells and transfer energy between the two – warm air from the tropics and cold air from the poles. At 30°N or 30°S of the Equator, air moves towards the poles across the Earth's surface until it meets the cold polar air at 60°N or 60°S. Here, the cold air pushes the warm air upwards and it cools. In the upper atmosphere, the air either moves to the poles or returns towards 30°N or 30°S, where it meets the descending air from the Hadley cells and returns to the Earth's surface.

The Hadley and polar cells are controlled by heat or cold and are known as thermally direct cells. The Ferrel cells are thermally indirect because they are controlled by the other two types of cell.

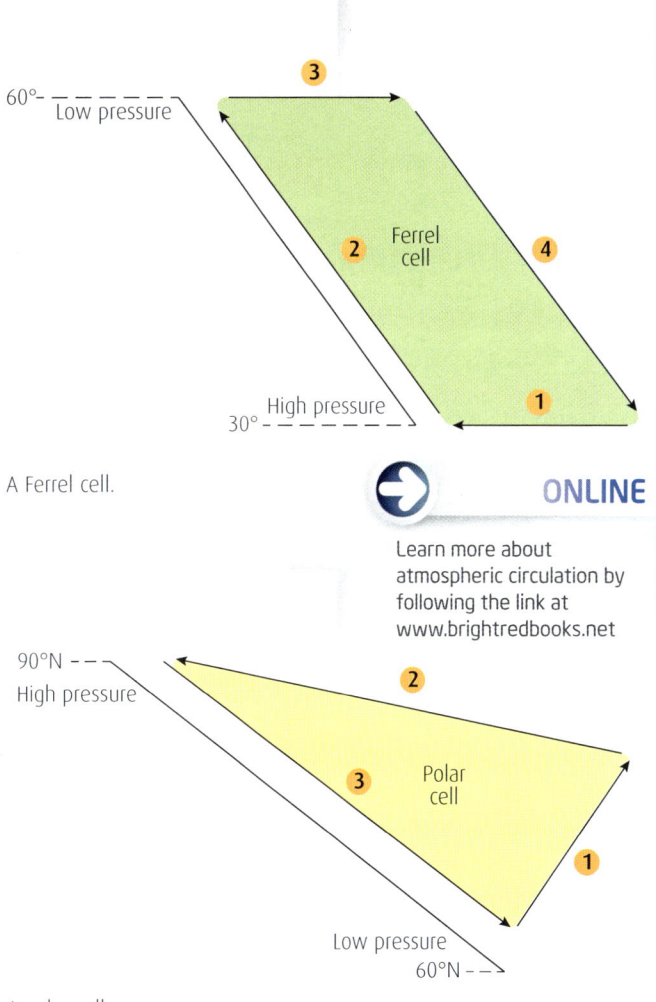

A Ferrel cell.

ONLINE

Learn more about atmospheric circulation by following the link at www.brightredbooks.net

POLAR CELLS

At the poles, cold air, which is heavier than warm air, sinks to form an area of high pressure. The cold polar air moves southwards across the Earth to about 60°N or 60°S of the Equator. At this point, the cold air meets warmer air and is pushed upwards. The air moves back towards the poles in the upper atmosphere to complete the polar cell.

A polar cell.

THINGS TO DO AND THINK ABOUT

Practise drawing the sketches of polar, Hadley and Ferrel cells.

Explain how circulation cells in the atmosphere and the associated surface winds assist in the transfer of energy between areas of surplus and deficit heat.

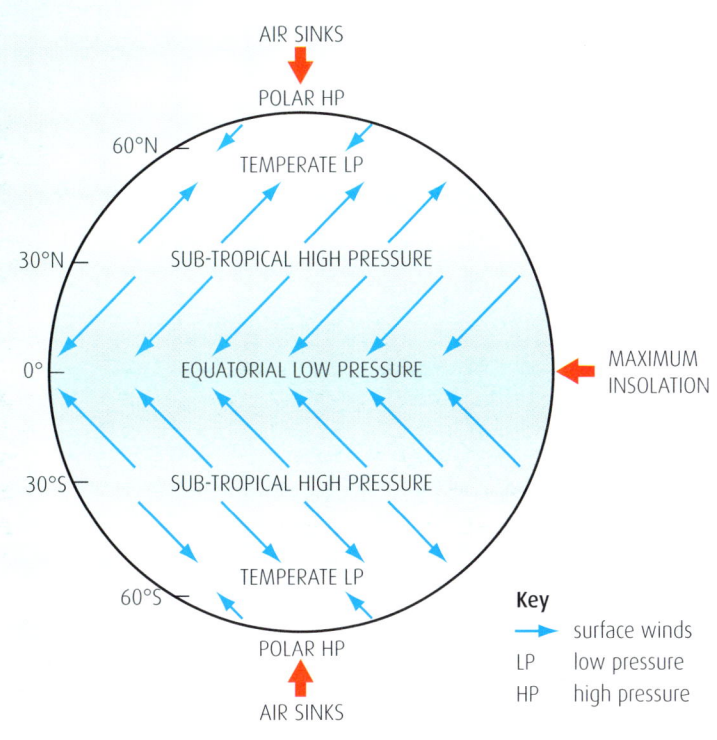

Key
→ surface winds
LP low pressure
HP high pressure

ONLINE TEST

Test your knowledge of energy transfer online at www.brightredbooks.net

DON'T FORGET

The three-cell model is a very simple way of explaining energy transfer. You are expected to be able to describe and explain this process in the exam.

ONLINE

Head to www.brightredbooks.net to see the answers for this question.

ENERGY TRANSFER 2

High pressure

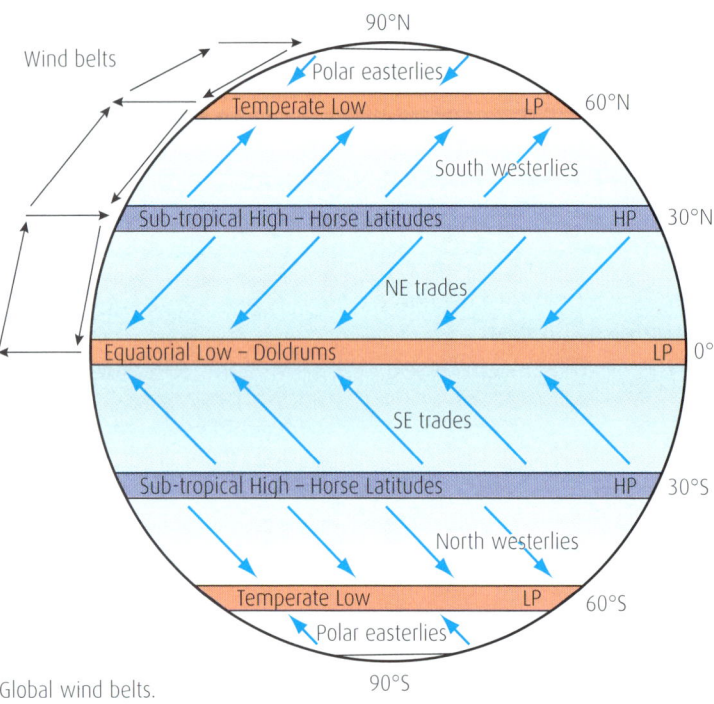

Coriolis effect

wind

pressure
gradient
force

Low pressure

The Coriolis effect.

GLOBAL WIND PATTERNS

The three-cell model shows surface winds moving energy from areas of high to low pressure. Winds always blow from high to low pressure and are named after the direction they are blowing from.

Coriolis effect

The Coriolis effect is caused by the Earth's rotation. In the Northern Hemisphere, the Coriolis effect deflects movement to the right. In the Southern Hemisphere, the Coriolis effect deflects movement to the left. The combination of the atmospheric cells and the Coriolis effect lead to the wind belts. The wind belts drive surface ocean circulation.

Wind belts

90°N
Polar easterlies
Temperate Low   LP   60°N
South westerlies
Sub-tropical High – Horse Latitudes   HP   30°N
NE trades
Equatorial Low – Doldrums   LP   0°
SE trades
Sub-tropical High – Horse Latitudes   HP   30°S
North westerlies
Temperate Low   LP   60°S
Polar easterlies
90°S

Global wind belts.

In the Northern Hemisphere:

- **polar easterlies** blow from the high pressure zone at the North Pole to about 60°N
- **westerlies** blow between 30°N to about 60°N – this area is affected by depressions and anticyclones
- the **north-east trades** blow from 30°N to the low pressure zone at the Equator.

In the Southern Hemisphere:

- **polar easterlies** blow from the high pressure zone at the South Pole to about 60°S
- **westerlies** blow between 30°S to about 60°S
- the **south-east trades** blow from 30°S to the low pressure zone at the Equator.

This is a simple model – in reality, the winds in the Northern Hemisphere are affected by the continents, for example, the Himalaya Mountains affect the monsoon climate and rains in the Indian Ocean.

OCEANIC CIRCULATION

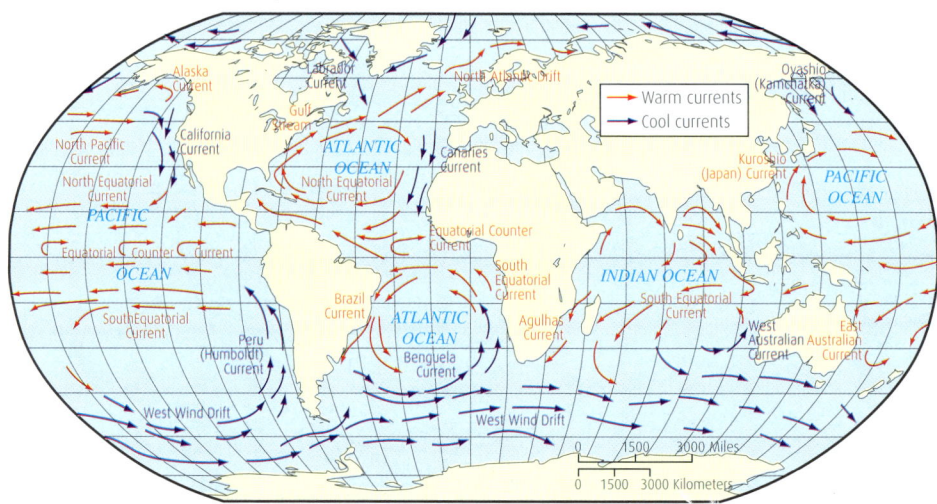

World ocean currents. Warm ocean currents are shown in red and cold ocean currents are shown in blue.

contd

The movement of the water in the oceans is closely connected to atmospheric circulation. The oceans move surplus energy from the tropics towards the poles. Warm water moves north or south towards the poles, while colder water moves from the poles to the Equator.

The Earth's rotation influences the movement of the oceans. As the Earth rotates from west to east, the oceans move in the opposite direction from east to west. Continents and other land masses deflect the movement of ocean currents.

Northern Atlantic Gyre

Giant circular movements of water can be identified in the Atlantic and Pacific oceans. These are known as **gyres**. Trace the following description of the Northern Atlantic Gyre onto the map on page 34.

In the Atlantic Ocean at the Equator, the **North Equatorial Current** moves west towards South America, where it is either deflected north towards the Caribbean Sea or south along the South American Coast as the warm **Brazilian Current**. The northerly movement of warm water flows through the Caribbean and into the Gulf of Mexico and up the coast of the USA as the warm **Gulf Stream**. It then crosses the Atlantic Ocean as the **North Atlantic Drift** towards the Arctic Ocean. To balance this movement north of warm water, cold currents carry water from the Arctic, for example, the **Labrador Current**. On the eastern side of the Atlantic along the African coast, the cold **Canaries Current** replaces the warm waters that flow west.

Patterns in the South Atlantic and Pacific oceans

In the Pacific, warm water at the Equator moves east towards Asia and Australia. When it reaches these land masses, it is deflected north or south. In the northern Pacific, the **Japanese Current** carries the warm water north towards Alaska. Cold currents flow down the eastern coast of North America, for example, the **Californian Current**.

As with the Atlantic, a mirrored pattern is seen in the southern Pacific Ocean.

This pattern of oceanic currents has strange effects.

- The west coasts of continents below latitudes of about 30° come into contact with cold currents, whereas the east coasts are in contact with warm currents. This helps to create the Earth's desert climates.

- The situation is reversed above latitude 45°, with the west coasts in contact with warm currents and the east coasts in contact with cold currents. This keeps the UK climate very mild.

The phenomena known as **El Niño** and **La Niña** – periodic warming and cooling of the tropical Pacific Ocean – also influence global weather patterns, causing extreme events such as heavy rainfall in desert areas.

 ## THINGS TO DO AND THINK ABOUT

Look up climatic information for the UK and Tomsk in Russia. They are at similar latitudes and yet the winter temperatures are very different.

Study the reference map (right), which shows selected ocean currents in the North Atlantic Ocean.

(a) Describe the pattern of ocean currents in the North Atlantic Ocean.

(b) Explain how they help to maintain the global energy balance.

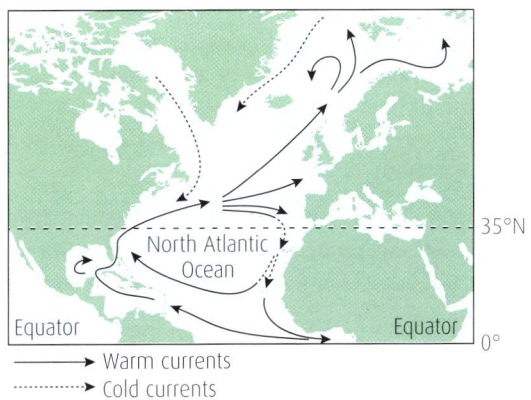

North Atlantic Ocean

Equator

Equator

35°N

0°

→ Warm currents
·····→ Cold currents

 **DON'T FORGET**

You need to be able to describe the pattern of the Northern Atlantic Gyre in detail.

 ONLINE TEST

Test your knowledge of energy transfer online at www.brightredbooks.net

 VIDEO LINK

Watch the clip at www.brightredbooks.net to learn how changes in ocean currents can lead to catastrophic weather conditions.

 ONLINE

Head to www.brightredbooks.net to see the answers for these questions.

CLIMATE CASE STUDY: WEST AFRICA – THE INTER-TROPICAL CONVERGENCE ZONE (ITCZ)

DON'T FORGET

The impact of the ITCZ links directly with the sections on rural land degradation in the Sahel region; see pp. 69–73.

AIR MASSES AFFECTING WEST AFRICA

Air masses are large masses of air that settle over areas of the Earth's surface. They take on the characteristics of the area over which they settle – the source area – and take these characteristics with them when they move to another part of the Earth's surface.

In Western Africa, the tropics, there are two main air masses:

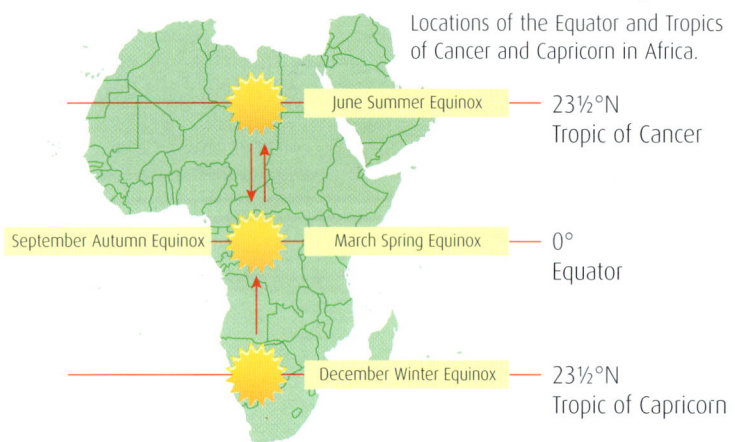

Locations of the Equator and Tropics of Cancer and Capricorn in Africa.

June Summer Equinox — 23½°N Tropic of Cancer

September Autumn Equinox — March Spring Equinox — 0° Equator

December Winter Equinox — 23½°N Tropic of Capricorn

- tropical maritime
- tropical continental

These air masses have their own characteristics and influence the areas with which they come into contact.

Tropical maritime (mT) air comes from the tropics and blows over water. It is a warm, wet wind and brings unstable conditions with high temperatures and heavy precipitation.

Tropical continental (cT) air also comes from the tropics, but blows over land. It is warm and dry and brings stable air, producing high temperatures and drought conditions.

The location of the ITCZ varies throughout the year
The ITCZ over land moves farther north or south than the ITCZ over the oceans due to the variation in land temperatures.

ITCZ JULY

ITCZ JANUARY

Changes in the seasonal location of the ITCZ.

When the mT air mass dominates as a result of its migration north, rain is also brought north to the Sahel desert region. When cT air dominates, West Africa experiences a dry season. The ITCZ causes this variation in rainfall. This has a direct effect on farming in the Sahel region. If the rains fail, then the crops fail. This links directly with the section on rural land degradation later in this study guide.

The North Pole tilts away from the Sun in December and tilts towards the Sun in June. This means that the North Pole experiences 24 hours of darkness in winter and 24 hours of daylight in summer. Everywhere on Earth has 12 hours of daylight and 12 hours of darkness in March and September – these are known as the **Spring** and **Autumn Equinoxes**.

We learned earlier that the Sun's rays are concentrated on the tropics and, in particular, on the Equator. This heat from the Sun warms the air near the Earth's surface, which then rises and starts the process of **atmospheric circulation**. If the Sun moves its focus north or south of the Equator, then the point at which the two Hadley cells in the tropics meet will also move. This is known as the **thermal Equator** and is responsible for the seasons experienced in each hemisphere.

The point at which the surface winds in these two cells meets is known as the **Inter-Tropical Convergence Zone (ITCZ)**. This is where concentrated energy from the Sun warms the air, which then rises into the upper atmosphere.

The ITCZ has a direct influence on the climate of West Africa.

Temperatures in this area are about 23°C all year. You can see from the map and bar charts on the next page that the amount of rainfall decreases moving north from the coast.

contd

Between the Equator and around 10°N, the bar chart for Abidjan shows that there is high rainfall all year round. This is associated with the mT air pushing its way north with the ITCZ throughout the first six months of the year.

In Ouagadougou, there is less rainfall throughout the year. The peak is linked with the path of the ITCZ, bringing the rainy season associated with the mT air as it moves inland.

In Nioro, the climate is dominated by the cT air – it has a very dry desert climate. The low rainfall is associated with the maximum northerly extent of the ITCZ and its mT air. This is part of the Sahel region and has highly variable rainfall.

As the ITCZ moves south, the amount of rainfall decreases. There is a twin peak in Abidjan as the mT air passes over this area twice, moving with the ITCZ.

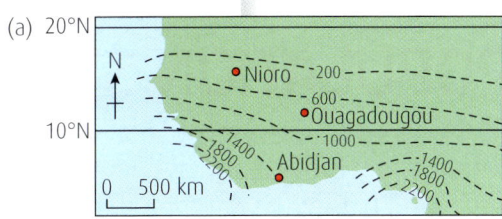
(a)

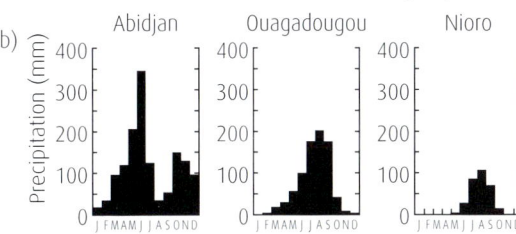
(b)

Rainfall in West Africa.
(a) Map showing average rainfall;
(b) bar charts of rainfall in selected locations.

- The maximum point north of the ITCZ – when the thermal Equator is over the Tropic of Cancer – occurs in July and brings some rain to continental Africa. The rain is associated with mT air that has pushed its way inland.
- The maximum point south of the ITCZ – when the thermal Equator is over the Tropic of Capricorn – occurs in January and brings the dry season to continental Africa. This is associated with cT air that has pushed its way south, dominating the continent.

ONLINE

Follow the link at www.brightredbooks.net to learn more about the ITCZ.

ONLINE TEST

Head to www.brightredbooks.net and test your knowledge of this section.

ONLINE

Head to www.brightredbooks.net to see the answers for these questions.

THINGS TO DO AND THINK ABOUT

1 Describe the origin, nature and characteristics of the tropical maritime and tropical continental air masses.

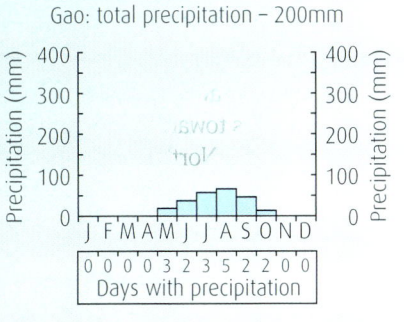
Gao: total precipitation – 200mm

	J	F	M	A	M	J	J	A	S	O	N	D
Days with precipitation	0	0	0	0	3	2	3	5	2	2	0	0

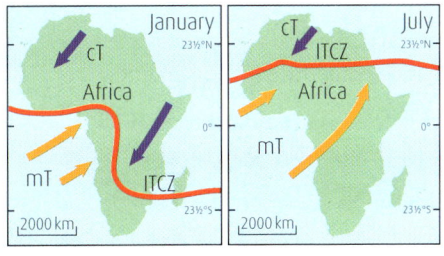

Key mT = Tropical maritime
  cT = Tropical continental
  ITCZ = Inter Tropical Convergence Zone

Location of selected air masses and the ITCZ in Africa in January and July.

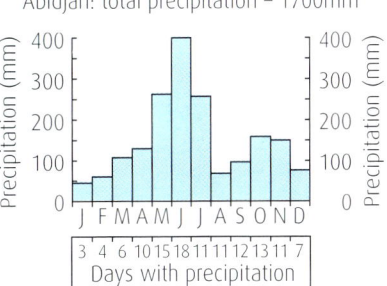
Bobo-Dioulaso: total precipitation – 1000mm

	J	F	M	A	M	J	J	A	S	O	N	D
Days with precipitation	0	1	3	4	7	9	12	16	12	5	2	0

Abidjan: total precipitation – 1700mm

	J	F	M	A	M	J	J	A	S	O	N	D
Days with precipitation	3	4	6	10	15	18	11	11	12	13	11	7

Average monthly rainfall/days with precipitation for three locations in West Africa.

2 Describe the rainfall patterns in West Africa.

3 Explain these patterns.

For the exam, you need to know the following for the Atmosphere topic:
- redistribution of energy by atmospheric and oceanic circulation
- causes, characteristics and impact of the ITCZ

POPULATION STRUCTURE

Demography is the study of population. Within this topic, you need to be aware of a wide range of terms and be able to describe different types of graphs. You will also be expected to explain differences between graphs, principally in relation to developed and developing countries. You may also be asked to explain trends on population graphs. You may be asked to interpret information in tables showing different measurements of population for two or more countries. Population growth and population decline can cause a variety of difficulties in countries and many governments have to put measures in place to accommodate such changes. Population planning is essential, whether in a developed or developing country.

POPULATION TERMS

Demography is the statistical study of populations.

The **crude birth rate (CBR)** is the number of births per 1000 people in a given population. The CBR does not take age and sex into account.

The **crude death rate (CDR)** is the number of deaths per 1000 people in a given population.

The **natural increase** is the difference between the CBR and the CDR and is usually quoted as a percentage per year. It indicates an increase in population per 1000 people.

A **natural decrease** occurs when a population is decreasing as a result of an excess of deaths compared with births. It is also a percentage and indicates a decrease in the population per 1000 people.

Life expectancy is the average number of years a person born in a particular country is expected to live.

The **infant mortality rate** is the number of infants under the age of 1 year old who die per 1000 live births.

The **literacy rate** is the percentage of the total population who can read and write.

The **gross national product (GNP)** is the total value of all goods and services produced by a country per year, divided by the total population of that country.

Migration is the movement of people into/out of/within countries.

A **pro-natal** policy encourages the growth and success of larger families, often through rewards and financial benefits.

An **anti-natal** policy discourages childbirth.

DON'T FORGET

Neither natural increase nor natural decrease take migration into account.

POPULATION PYRAMIDS

The population structure of a town, region or country is the number of people of male and female sex in different age groups. **Population pyramids** are used to display this age/sex structure and can be studied and compared. They are displayed as bar charts drawn back-to-back to make comparisons between age and sex easy to identify.

Population pyramids show:

- the percentage of men/boys and women/girls in each age category
- men/boys are shown on the left and women/girls on the right
- the percentage of the population in each five-year age bracket
- the percentage of people in the active and dependent populations.

contd

Population pyramids are useful in predicting future population changes and planning ahead for these changes. For example, governments and local authorities use them to plan the development or closure of facilities such as educational centres and health centres/services.

Countries around the world are all at different stages of development and population pyramids highlight the associated characteristics of these stages in their shapes.

Population pyramids for (top) Germany (developed country) and (bottom) Tanzania (developing country).

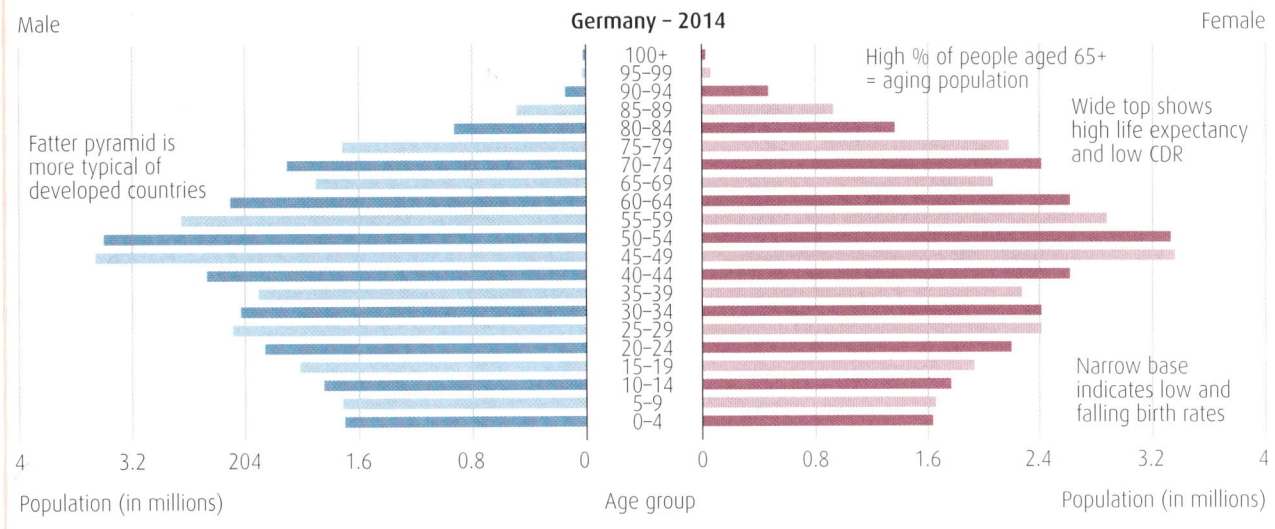

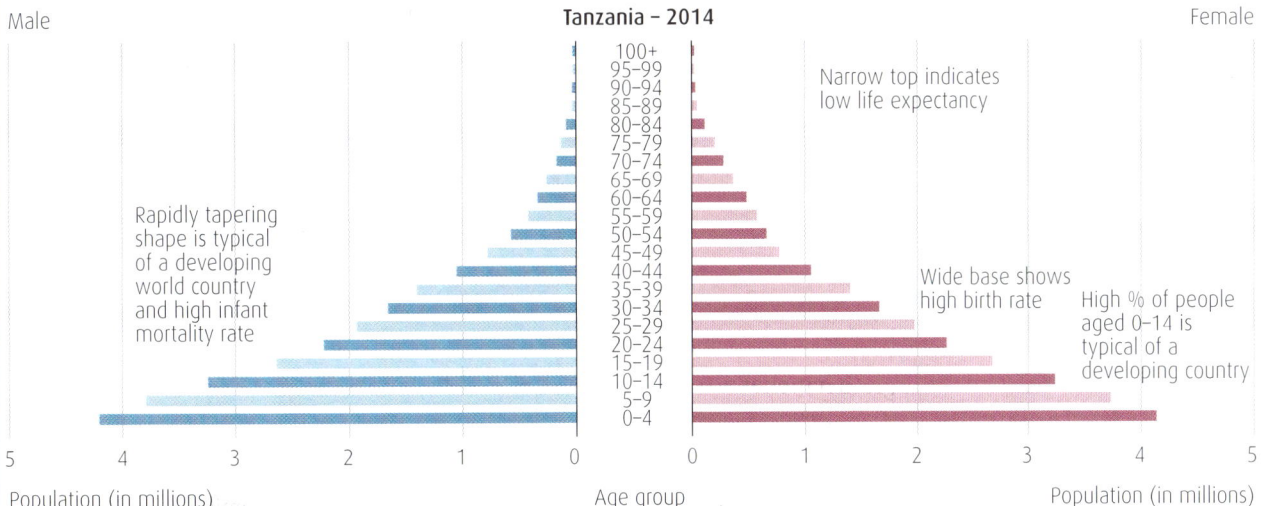

Both structures have their benefits and disadvantages for populations. They both hold key information about the country, in particular the proportion of the population that is **active** or **dependent**. Each population structure holds clues to the state of development of the population.

THINGS TO DO AND THINK ABOUT

Population pyramids are a great tool to make quick comparisons between the population and development of countries. They even allow you to suggest future concerns for the country – for example, the associated problems of an ageing population and a low birth rate.

 ONLINE

You can check out the population pyramid for every country in the world by following the population pyramid link at www.brightredbooks.net

 ONLINE TEST

Take the test on population pyramids at www.brightredbooks.net

DEMOGRAPHIC TRANSITION

DEMOGRAPHIC TRANSITION MODEL

The demographic transition model indicates changes in birth and death rates and, consequently, changes in the population over time. The model has five separate stages and highlights how the CBR and CDR are influenced by:

- social factors
- economic factors

This is useful in understanding the changes in global population with time. As social and economic factors affect the model directly, it allows us to make predictions about the future of a population and its size. However, the model does not take migration into account.

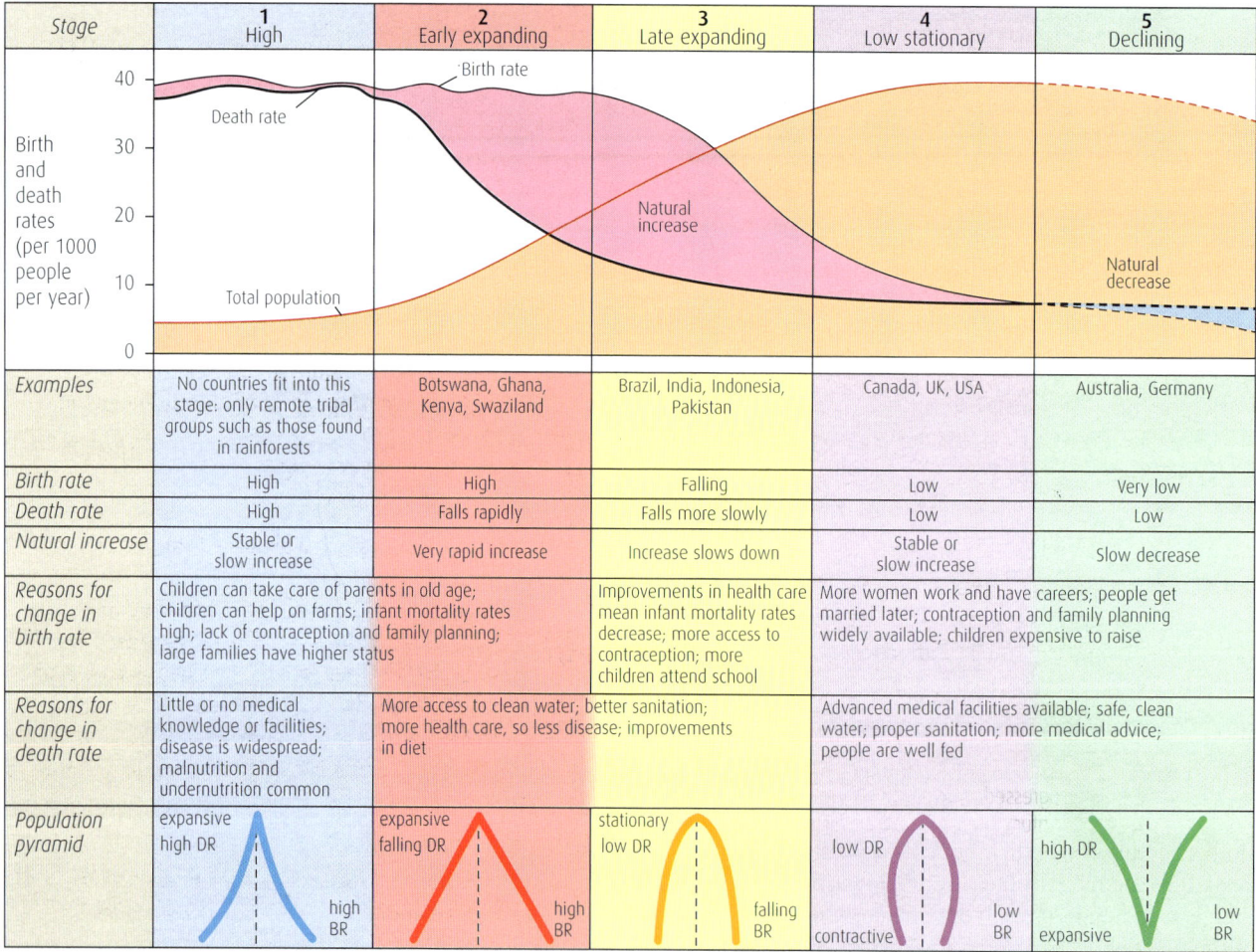

Stage	1 High	2 Early expanding	3 Late expanding	4 Low stationary	5 Declining
Examples	No countries fit into this stage: only remote tribal groups such as those found in rainforests	Botswana, Ghana, Kenya, Swaziland	Brazil, India, Indonesia, Pakistan	Canada, UK, USA	Australia, Germany
Birth rate	High	High	Falling	Low	Very low
Death rate	High	Falls rapidly	Falls more slowly	Low	Low
Natural increase	Stable or slow increase	Very rapid increase	Increase slows down	Stable or slow increase	Slow decrease
Reasons for change in birth rate	Children can take care of parents in old age; children can help on farms; infant mortality rates high; lack of contraception and family planning; large families have higher status		Improvements in health care mean infant mortality rates decrease; more access to contraception; more children attend school	More women work and have careers; people get married later; contraception and family planning widely available; children expensive to raise	
Reasons for change in death rate	Little or no medical knowledge or facilities; disease is widespread; malnutrition and undernutrition common	More access to clean water; better sanitation; more health care, so less disease; improvements in diet		Advanced medical facilities available; safe, clean water; proper sanitation; more medical advice; people are well fed	
Population pyramid	expansive high DR / high BR	expansive falling DR / high BR	stationary low DR / falling BR	low DR / low BR contractive	high DR / low BR expansive

Stage one: high fluctuating

Features of this stage are:
- birth rate and death rate are both high and both fluctuate, but the death rate fluctuates slightly more (e.g. as a result of famine)
- growth is restricted
- large family sizes
- high infant mortality rate
- low life expectancy
- little population change
- total population is very low
- natural increase is very low.

The reasons for this are:
- children can take care of the elderly and sick
- children can work on farms and create alternative income for the family making them an economic asset
- lack of contraception
- lack of sex education
- large families have status
- little or no medical/nutritional knowledge.

Example

Scotland was in stage one before 1760, but today this pattern is only seen in a few of the least developed areas of the world, for example, remote tribes in the Amazon rainforest and New Guinea.

contd

Stage two: early expanding

Features of this stage are:
- dramatic drop in death rate
- birth rate remains high
- reduced infant mortality rate
- total population increasing
- natural increase is high.

The reasons for this are similar to stage one and also include:
- introduction of clean water supplies, sewage systems and health care systems
- people have not yet realised that traditionally large families are not required as a result of the lower infant mortality rate.

Example

Scotland was at this stage between 1760 and 1870. Most countries have now left this stage as a result of improved health care. Current examples include Kenya and Swaziland.

Stage three: late expanding

Features of this stage are:
- the birth rate begins to drop, but remains higher than the death rate
- death rate continues to fall
- rapidly increasing population, although rate is beginning to slow compared with stages one and two
- infant mortality rate falls
- natural increase is fairly high.

The reasons for this are:
- introduction of clean water supplies
- access to family planning and contraception
- improvements in health care
- improvements in living standards
- improvements in diet
- better sanitation.

Example

Many countries progressed through this stage at the beginning of the twentieth century. Scotland was at this stage from 1870 to 1950. Countries currently in this stage include India and Brazil.

Stage four: low stationary

Features of this stage are:
- both birth and death rates are low as the decrease in birth rate has caught up with the death rate
- total population is steady
- natural increase is very low.

The reasons for this are:
- more women working and putting careers first
- marriages later in life, if at all
- contraception and family planning widely available
- children are seen as an economic drain
- advances in medical facilities and health care
- a good standard of sanitation has been achieved
- medical advice readily available
- nutritional advice available.

Example

Scotland entered this stage in 1950 and remains at this stage. Developed countries such as Canada and the UK are fast approaching this stage. Most developed countries are at this stage now.

Stage five: declining population

Features of this stage are:
- birth rate continues to fall
- death rate remains low, but may increase slightly as a result of the large proportion of elderly people in the population
- total population is decreasing
- natural increase is negative.

The reasons for this are similar to stage four and also include:
- couples are having children later in life, if at all
- an ageing population structure leads to a need for more care home facilities, specialised health care and a need for subsidies, for example, bus passes and prescriptions – this puts pressure on a decreasing active population to provide sufficient funding in the form of taxes.

Example

Only a handful of highly developed countries are currently in this stage, for example, Germany and Australia. Scotland is progressing quickly towards this stage.

 THINGS TO DO AND THINK ABOUT

Every country can be shown to move through the demographic transition model. It has, however, been criticised as it is largely based on European history. Some aspects have not been considered in the model, such as rural and urban variations. You should always try to refer to case studies when using the demographic transition model.

CENSUS

COLLECTING AND COLLATING DEMOGRAPHIC DATA

There are several ways of collecting population data and each method has some limitations. These methods include:

- censuses
- civil registration – births, marriages and deaths
- national, trans-national and global surveys.

Value of data collection

A variety of demographic sources allows general conclusions to be drawn about the social and economic state of a population. However, data collection is extremely costly and challenging. Censuses were originally carried out for taxation purposes and for army recruitment. Today, information from censuses is mainly used to plan for provisions such as:

- maternity care
- nursery, primary and secondary education
- higher education
- employment, housing and transport
- pensions
- geriatric health care and sheltered housing.

Censuses

Population data is collected through a **national census**. Many countries use a detailed form to obtain information about the social and economic characteristics of a population, and the collection of census data has a long history. Most countries collect data on their population.

The data collected includes information on every member of a household, including their **age**, **sex**, **religion**, **health**, **job**, **education**, **birth place** and their **means of travel/transport to work**. The census also collects information on **living conditions**, **house tenure** and **amenities**. In **Scotland**, a question about knowledge of the **Gaelic language** is included in the census and, in **Wales**, a question on **Welsh language** has been asked since 1891.

A census is carried out every 10 years in the UK. The collection of census data is the responsibility of the three UK Census Offices – The Office of National Statistics in England and Wales, National Records of Scotland and the Northern Ireland Statistics and Research Agency. The census gives a snapshot of social and economic development.

Census questionnaires are posted out and residents can opt to complete the paper questionnaire and return by post or complete it online. The results are collated and analysed to interpret changes in population, to forecast population trends, to plan for future demands on services and for strategic planning. Censuses also allow governments to make informed decisions about whether to encourage or discourage births and migration. Currently, this is not the case in the UK as our growth rate is very low.

Difficulties in collecting demographic data

Demographic data collection is continually improving in terms of quality and quantity. The collection of census data is more reliable in countries that are more developed economically, although this is not always the case. For example, people may be unwilling to complete a census for fear of increased taxation. In the UK in 1991, over one million people did not complete the census.

The distortion of census data is related to a combination of factors. Developing countries face a variety of obstacles to data collection:

ONLINE

Learn more about Scotland's censuses online at www.brightredbooks.net

contd

- nomadic populations are difficult to include
- urban–rural migration
- poor communication links, difficult terrain and scattered settlements
- low literacy rates
- a variety of languages – it is expensive to translate a census into many languages and there are often unofficial languages/local dialects
- expense – debt-ridden countries have other priorities
- ethnic tensions, which can influence the accuracy of the data
- under-registration for social and religious reasons, for example, under recording of women because of the one-child policy in China
- civil wars leading to mass migration
- poor infrastructure

Example

Nigeria conducted a population census in 2006. However, the chair of the National Population Commission stated in 2012 that 'Nigeria has no data. People cannot really tell you precisely what the population is'. Another census will be conducted in 2016.

 ## THINGS TO DO AND THINK ABOUT

Look at the question and answer below.

Q. Explain the problems of collecting accurate population data in developing countries.

 ### ONLINE TEST

Test yourself on this topic online at www.brightredbooks.net

A. Migration can have a detrimental impact on collecting accurate population data because large numbers of migrants, e.g. the *Tuareg* or *Fulani* in West Africa and the shifting cultivators of the Amazon, may lead to people being missed or counted twice, making the statistics flawed. Countries with large numbers of homeless people or large numbers of people who migrate from rural to urban areas and live in shanty towns, e.g. Makoko in Lagos, Nigeria, will have no official address for an enumerator to visit. Poor communication links and difficult terrain, e.g. in the Amazon rainforest, make it difficult for enumerators to reach isolated villages. The variety of languages spoken in many countries (over 500 in Nigeria) make it difficult to provide forms that everyone can complete, as it would be extremely costly to translate the form into over 500 languages; as a result completing census data is inaccessible for many. Also, the considerable costs involved in printing, training enumerators, distributing forms and analysing the results can make conducting a census impossible, especially when the country may have more pressing problems such as housing and education to address, which will ultimately help to improve living conditions and development. In countries with a high level of illiteracy, mistakes may be made and more enumerators will be needed to help, which is extremely costly. People may be suspicious of why the census is being conducted, and may lie in an attempt to protect themselves. Ethnic tensions and internal political rivalries may lead to inaccuracies, e.g. northern Nigeria was reported to have inflated its population figures to secure increased political representation. Under-registration may occur for social, religious and political reasons, e.g. China's one-child policy may have reduced the registration of baby girls and therefore many may not be registered to complete the form. In countries suffering from war, e.g. Afghanistan, it may be dangerous for enumerators to enter regions and data will quickly become outdated.

Try writing your own answer to this question using examples you have studied in class.

MIGRATION

When individuals move from one place to another, on either a permanent or semi-permanent basis, this is called **migration.** This can be for periods of years, months, or be seasonal migration.

TYPES OF MIGRATION

Migration may be temporary or permanent, voluntary or forced. Migration is also categorised by the reason for movement.

- **Refugee** – a refugee is someone who is stateless. There are a number of ways for someone to become a refugee, including war, religious or ethnic persecution, or famine. Refugees often do not have many belongings or any documentation to prove who they are.
- **Asylum seeker** – an asylum seeker is someone who is looking for shelter because they have found themselves in a threatening situation in their home country. They must declare this when they arrive in their host country. Asylum seekers may be fleeing religious or ethnic persecution and feel that they are in immediate or real danger of imprisonment or death if they were to return to their homeland.
- **Economic migrant** – an economic migrant is someone who has made the decision to move to earn more money and to directly improve their quality of life. Often, economic migrants send remittance payments back to their families in their home country.
- **Permanent** – long-term migration.
- **Voluntary** – migration when people have made a free and conscious decision to move.
- **Forced** – this is when individuals have no choice but to migrate.
- **International** – this involves crossing a national boundary where identification checks are performed.
- **Internal or intra-national** – this is when the migration occurs within a country or even within a city.

There are basic factors that encourage individuals to migrate from one area to another. These can be categorised into push and pull factors that either encourage people to leave their country of origin or attract people to a new location. Sometimes these basic factors are aspirational and may not become a reality when the move has been made.

DON'T FORGET

Migration brings both **advantages** and **disadvantages** for the **source location** and the **new location**.

ONLINE

Look at the link at www.brightredbooks.net and makes notes under the following headings:
- what is migration?
- types of migration
- people who migrate
- impact of migration

RURAL–URBAN MIGRATION

Developing countries tend to have a large number of individuals moving from the countryside to the cities or larger towns. **Push** factors encourage people to move away from rural areas and **pull** factors attract them to cities.

Pull Factors		
1 More jobs people expect to find better paid jobs	**2 Better education** more schools and colleges	**3 Better health care** more doctors, hospitals
4 Amenities large shops, cinemas	**5 More houses** people expect to find better housing	

Push Factors		
1 Not enough land to farm because of growing population	**2 Crop failure** because of disease, lack of rain and soil	**3 Poverty** most farmers grow only enough to feed themselves (subsistence farmers); they do not sell anything, so they have no money
4 Natural disasters flood, hurricanes, earthquakes, drought	**5 Farm machines** mean fewer jobs for farm workers	

Push and pull factors in migration.

Push factors include:
- not enough land
- crop failure
- poverty and low living standards
- natural disasters
- mechanisation of farming

Pull factors include:
- greater availability and variety of jobs
- better education
- better health care facilities
- more amenities
- more houses

contd

Effects of migration on the countryside

Rural to urban migration affects the **population structure** of rural areas. Often, the active male population moves away, leaving behind very few young adults. As a result, the countryside loses its most active population, that is, those who can do the most work and have the most ideas. This means the farming is carried out by middle-aged and older people, especially women. There is often a definite sex and skills imbalance in rural areas, and these areas suffer as a result. However, migration out of rural areas also means there is a reduced pressure on the land and on the limited resources available for housing.

Effects of migration on the city

The growth of cities is called **urbanisation**. Often, developing cities simply cannot cope with the vast number of immigrants arriving in search of a better quality of life. When immigrants arrive, the stark reality of the city is often not what they expected. **Housing problems** are a direct effect of a high volume of immigrants. There are not enough houses for everyone, so people build makeshift houses on waste land. These squatter settlements are called **shanty towns** and many people who move from rural to urban areas end up living in them. There are not enough full-time jobs for everyone and both adults and children turn to **informal jobs** as their only source of income. These include selling fruit, collecting rubbish or running errands. This is an unreliable source of income and is often not enough to live on. **Environmental problems** (for example, air pollution and a lack of clean water, sewage treatment and rubbish disposal facilities) are often a direct result of the poor quality of dwellings in the shanty town and poor sanitation; disease is widespread as most people cannot afford health care or enough food.

Very few regular jobs so adults and children take informal jobs, e.g. run errands, beg, collect rubbish.

No roads.

Disease and malnutrition very common.

'This shanty town is illegal. It is also unhealthy. We are going to bulldoze it.' (City Official)

'But if you do that we will just find another part of town to build our houses.' (Shanty Town dweller)

No proper sewage disposal, no clean water, no schools or clinics.

Houses made out of scrap wood, cardboard, tin, cloth.

No electricity.

Problems of shanty towns.

Combating rural-urban migration

Developing countries have tried to solve the problems caused by rural to urban migration in three main ways:

- improvements to shanty towns
- schemes for sites and services
- improving life in the countryside

Shanty towns are often bulldozed as they are illegal. This temporary fix does not last and it has detrimental effects. Individuals lose what little they have and have to start again from scratch and the shanty town is often rebuilt within a matter of days. One solution is to make these settlements legal. This has a positive effect as it makes it easier for the residents to gain employment. Access to **electricity** also reduces the amount of crime within the settlements and makes them safer places to live. Access to **clean water** and proper **sewage disposal** also reduces the spread of disease.

The authorities often struggle to provide suitable housing for everyone, hence the growth of shanty towns. One solution to this is to designate areas where basic services are provided. People are often allowed to build their own houses in these areas with money borrowed directly from the authorities.

Life in the countryside in developing countries is hard, but push factors in the countryside can be reduced to encourage people to stay. Improvements may include the development of **schools** and **health centres** and making sure that farming in the countryside is profitable and that local people have the **skills** and **training** to make it a success.

 THINGS TO DO AND THINK ABOUT

Refer to case studies as much as possible when discussing the reasons why people migrate and what impact this has on the receiving area and the area of origin.

 **DON'T FORGET**

Migration occurs in both developing and developed countries. In developed countries, urban to rural migration is more prevalent, whereas in developing countries, rural to urban migration is more common.

 ONLINE

Learn more about migration by following the link at www.brightredbooks.net

 ONLINE TEST

Head to www.brightredbooks.net and test yourself on this topic.

CASE STUDY: VOLUNTARY MIGRATION

MIGRATION FROM EASTERN EUROPE TO THE UK

Voluntary migration often occurs when people want to improve their circumstances.

European Union

The **European Union** (EU) is an economic and political union of 28 countries. The EU operates an internal market that allows the free movement of goods, capital, services and people between member states to avoid conflict within European countries. This concept was created following the aftermath of the Second World War.

The current EU countries are: Austria, Belgium, Bulgaria, Croatia, Republic of Cyprus, the Czech Republic, Denmark, Estonia, Finland, France, Germany, Greece, Hungary, Ireland, Italy, Latvia, Lithuania, Luxembourg, Malta, the Netherlands, Poland, Portugal, Romania, Slovakia, Slovenia, Spain, Sweden and the UK.

Eastern European overview

The most recent country to join the EU was Croatia in 2013. Romania and Bulgaria joined in 2007. However, the largest expansion of the EU was in 2004 when Poland, Hungary and the Czech Republic joined. A large number of immigrants exercised their new-found right to move freely within the EU and **voluntarily** relocated to the UK in search of a higher standard of living and economic income.

The BBC has reported that approximately 600 000 Eastern European migrants settled in the UK between 2004 and 2006. Many Eastern European migrants found jobs in construction, hospitality, catering, administrative work, business and retail, helping to fill gaps in the UK's labour market. They often earn up to five times as much as they did in their homeland. Many migrants send **remittance payments** home to their families in Eastern Europe.

When the EU and UK economies began to struggle in 2008, many Eastern Europeans went home. This means that their migration was **temporary**.

CAUSES OF MIGRATION FROM EASTERN EUROPE

Eastern European **push** factors

- Jobs in homeland are poorly paid.
- Eastern Europe's economy is less developed.
- Lower standards of living in homeland.
- Little employment available for skilled workers.
- Limited opportunities created a diaspora.

UK **pull** factors

- Access to education and health care.
- Higher standard of living.
- Job opportunities, education, higher wages.
- UK citizens unwilling to take up low-paid employment.
- Remittance money to send home.

IMPACTS OF MIGRATION FROM EASTERN EUROPE

Advantages to Eastern Europe

- Fewer people means less pressure on limited resources.
- Fewer people means less unemployment.
- Birth rate is lowered as male migrants are away – generally the active population, aged 18–30, migrates, resulting in a decline in population.
- Migrants return with new skills ('brain gain') leading to a better standard of living.
- Remittance money sent home to families.

contd

Disadvantages to Eastern Europe

- The most educated and healthy leave, which may hinder development.
- Women are the sole carers and look after both children and land.
- The active population leave (aged 18–30 years), so the dependent population is left behind.

Advantages to the UK

- The UK receives tax from migrant workers.
- Jobs that British people will not do are filled, for example, vital jobs within the NHS, construction and farm work.
- Cultural and linguistic mix.
- Brain drain/brain gain – many Eastern European migrants are highly skilled workers and/or graduates.
- Cheap labour.
- Drives economy so businesses have increased profits.
- Surplus of work drives down cost of labour.

Disadvantages to the UK

- Some Eastern Europeans work informally and do not pay tax, meaning British firms are undercut.
- Added pressure on services such as health care and education.
- Language barriers.
- Cultural differences leading to racial tension.
- Brain drain/brain gain – many highly skilled workers and graduates return home to Eastern Europe after qualifying in UK.
- Housing system may become strained due to large number of migrants.
- Welfare system may become strained if migrants claim benefits.

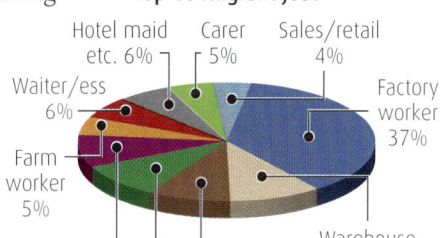

Top 10 migrant jobs

Ten most popular jobs held by EU migrants to the UK.

 ## THINGS TO DO AND THINK ABOUT

In the exam, population questions often refer to graphs, diagrams and statistics. You should make connections between areas of study so that you can draw on other knowledge that you may have to improve your exam response.

> Referring to a named case study, analyse the impact of migration on either the donor country or the receiving country.

The UK is more developed than many Eastern European countries, such as Poland, and, as a result, offers Eastern European workers many opportunities that they do not have in their homeland. The UK receives a large number of migrants from Eastern European countries that have joined the EU. The opportunities that the UK offer migrant workers include access to a higher standard of living and job opportunities. This voluntary migration brings many benefits for the receiving country. Migrant workers are willing to undertake jobs that British citizens are unwilling to complete and are often seen as cheap labour. This has closed a gap in the employment market as many migrant workers are now employed in vital roles within the NHS, the construction industry, driving buses and doing agricultural work. Other advantages include the UK benefitting from the prior graduate education and skills that migrant workers have. This creates a brain drain from Eastern European countries, where the most skilled workers move from the country where they trained to work elsewhere for improved pay and conditions. Utilising this skilled labour has driven the UK economy forward and businesses have made huge profits from migrant workers in the UK. These workers also pay tax to the UK economy, which has contributed to improved conditions and better services within the UK. Migrant workers also bring with them cultural and linguistic benefits, increasing diversity.

This model answer only includes the positive aspects of voluntary migration from Eastern Europe to the UK. Use the information in this section to help you list the disadvantages for the UK. Then do the same for the advantages and disadvantages for the country of origin.

 ## DON'T FORGET

You should think about the advantages and disadvantages for both the country of origin and the receiving country as there is an impact on both.

 ## ONLINE

Learn more about immigration in the UK by following the link at www.brightredbooks.net

 ## ONLINE TEST

Test yourself on migration online at www.brightredbooks.net

CASE STUDY: FORCED MIGRATION

CASE STUDY: MIGRATION FROM SYRIA

Forced migration can result from a range of circumstances, including sudden or life-threatening events such as war, famine or natural disasters.

It has been said that the Syrian civil war has led to the worst humanitarian crisis of our time.

Syria: key figures

- 12·2 million people in need within Syria.
- 7·6 million internally displaced people.
- 5·6 million children in Syria are estimated to be living in dire conditions including living in poverty, being displaced and being caught in the line of fire.
- 3·9 million Syrians have fled to neighbouring countries.
- 150 000+ people were killed between March 2011 and March 2014.
- The number of displaced children in Syria has more than tripled in the space of a year, from 920 000 to nearly 3 million.
- The Syrian people are now the largest refugee population in the world.

Syrian refugees forced to flee their homes.

Syria: overview

The Syrian crisis had been ongoing for some time when anti-government demonstrations began in March 2011. The attempt to prohibit these demonstrations led to the situation quickly escalating after the Syrian government carried out a violent crackdown on demonstrators; rebels then began to fight back against the regime.

Many defected from the army and began to assemble a Free Syria Army. Many civilians took up arms to join this cause. The situation was made worse by divisions between secular and Islamist fighters and there was also tension between ethnic groups, which continues to complicate the politics of the conflict. Many innocent civilians have been caught up in the conflict and civil war. There have been widespread violations of human rights and horrific mass murders have been reported through the bombing and destruction of major cities and towns. This has led to a situation where many people have to flee for their lives, resulting in **forced migration**.

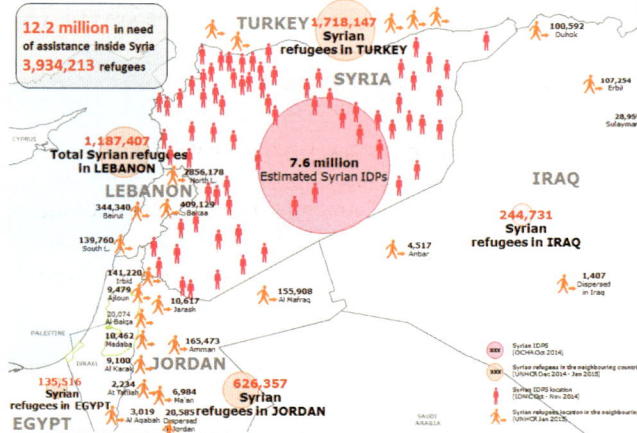

Numbers of refugees fleeing violence in Syria.

Many people are displaced within Syria and many refugees are also fleeing to neighbouring countries. This, in turn, has had a detrimental effect on the infrastructure of these countries as their resources come under extreme pressure to cope with the large numbers of migrants fleeing Syria.

SYRIAN CONFLICT: CAUSES

Push factors in Syria

Following the uprising and demonstrations, President Al-Assad promised to make changes and lifted the country's state of emergency law, which, for the previous 48 years, had given the regime the power to detain anyone without charge and to hold them indefinitely. However, just four days after the emergency law was lifted in April 2011, the Syrian regime sent out thousands of troops. This led to civil war, mass arrests, persecution, unrest, violence and fear for lives.

Pull factors of Lebanon

Factors that pulled immigration towards Lebanon include:

- safer environment
- access to health and education systems

contd

- Lebanon has always been a cross-roads for traders and is culturally diverse – it initially welcomed refugees from Syria
- strong economic growth
- cultural hub
- proximity – Lebanon and Syria share roughly 365 km of border.

IMPACTS OF SYRIAN CONFLICT

Advantages for Lebanon

The **advantages** for Lebanon were:
- arrival of aid agencies injected new money into the local economy
- creation of job opportunities
- local businesses benefited from the arrival of a supply of cheap labour
- landlords and landowners made significant profits from selling or renting land/properties
- cultural diversity – globalisation.

Disadvantages for Lebanon

Just over 1 million people have fled into Lebanon from Syria. The **disadvantages** for Lebanon have been:
- sharing of key resources – water and electricity
- strain on health and education, leading to epidemics such as scabies during the hot summer months
- overcrowding – some towns in the Bekaa Valley and the north have reported up to a 100% increase in population in the last two years
- economic impact
- crime/insecurity – such as prostitution, stealing
- Lebanese people are resentful of the Syrian workforce
- property prices are rising as a result of increased demand
- significant food price increases over the last 12 months, resulting from an increase in demand without additional supplies entering the market
- Lebanon itself suffers from political instability – the presence of such a large number of refugees threatens to tip Lebanon's fragile and volatile situation to breaking point
- many are living in tents with only basic sanitation and health care and dirty water – disease and malnutrition are spreading
- Lebanon has already welcomed many large influxes of refugees from countries such as Palestine – 450 000 Palestinian refugees live in Lebanon
- in 2006, parts of Lebanon's infrastructure were significantly damaged during the Hezbollah–Israel war – refugees add to this strain on resources
- the gap between the rich and poor is widening
- the conflict could spread across the border – rebels often smuggle weapons in and out of Syria through Lebanon
- friction between existing sectarian groups is worsening, for example, the Shia and Sunni Muslim communities
- on 22 March 2013, the Prime Minister of Lebanon resigned after deep-rooted unrest – this has left the country in turmoil and politically fragile
- child exploitation and child labour are increasing.

 THINGS TO DO AND THINK ABOUT

Consider the implication of forced migration on the receiving country and how its resources and infrastructure may be positively or negatively affected as a result.

Referring to a named case study, analyse the impact of migration on either the donor country or the receiving country.

For the exam, you will need to know the following for the Population section:
- methods and problems of data collection
- consequences of population structure
- causes and impacts of forced and voluntary migration

 ONLINE

Read more about immigration from Syria to Lebanon by following the link at www.brightredbooks.net

 DON'T FORGET

There are many examples of forced migration that you could refer to. The Syrian example is interesting because it has featured so much in the news and is a current event.

 ONLINE TEST

Take the test on migration online at www.brightredbooks.net

URBAN CHANGE

CATEGORIES OF URBAN CHANGE

Urban geography is concerned with the key aspects of cities and their change with time. Types of urban change that can occur in a city are shown in the diagram below and include housing, industry, leisure, transport and retail. Changes take place in urban areas within both developing and developed countries. This course focuses on changes in both **housing** and **transport** within urban areas. Cities are intensive hubs where the provision of accommodation and services for increasing numbers of residents is challenging.

Types of urban change that can occur in cities.

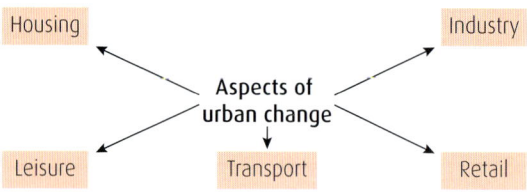

POPULATION GROWTH

In 2014 54% of the world's global population could be classified as **urban**. This has increased from only 34% in the 1960s. The urban population was previously concentrated in developed cities such as London and New York. However, urban growth is now concentrated in the less developed countries and regions of the globe such as São Paulo and Mexico City. Within developing countries in particular, there is a strong **rural to urban migration** and it is predicted that in the near future the vast majority of the population will live in urban areas as they try to achieve an improved standard of living.

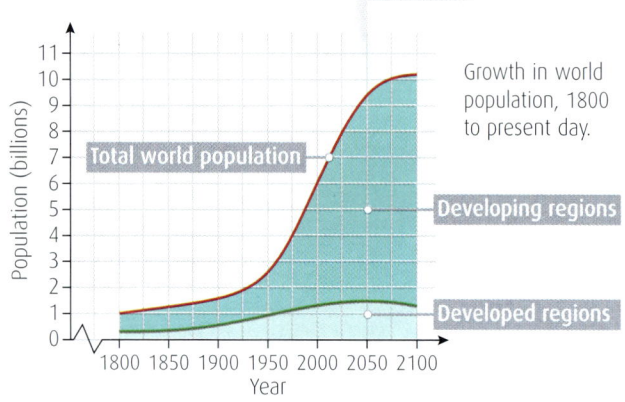

Growth in world population, 1800 to present day.

The first graph to the left highlights how the world's population has grown since the 1950s. This is mainly a result of improvements in sanitation and health care. Most of this growth has been in poorer or developing countries as these improvements have reached previously undeveloped areas.

The second graph shows that urban areas in developing countries are growing at a much faster rate than those in developed countries. This type of growth is expected to continue, with urban populations reaching record levels. This is linked with rural to urban migration.

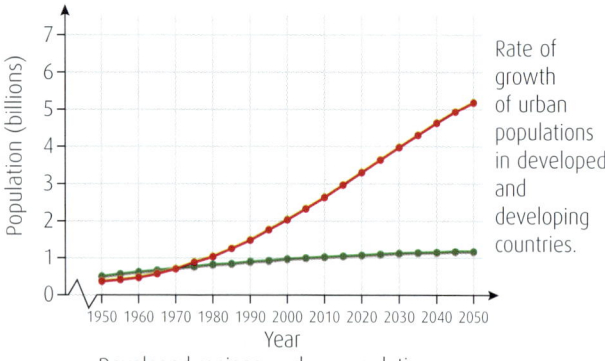

Rate of growth of urban populations in developed and developing countries.

- Developed regions – urban population
- Developing regions – urban population

REASONS FOR URBAN GROWTH IN DEVELOPING COUNTRIES

Reasons for this type of growth include:

- increasing birth rates
- decreasing death rates
- improved health care
- longer life expectancies
- employment opportunities

- better paid jobs
- migration patterns, for example, rural to urban
- migration of the active population to urban areas.

PROBLEMS IN URBAN AREAS

Rapid and increased population growth has led to many problems in urban areas.

Traffic

One of the main problems associated with population growth in both developing and developed countries is **traffic congestion.** Traffic congestion has become a major problem with the continued increase in car ownership. Many people wish to take their car to work for convenience. However, the city centre is also a route centre, where traffic congestion is compounded by large numbers of people entering the city on public transport and in privately owned cars via roads and motorways. This often causes **bottlenecks** in the central business district (CBD), where roads are narrow and there are many junctions where traffic is continually stopping and starting.

In developed countries, traffic congestion is made worse as people move to the edges of cities and **commute** into the city centre.

Other traffic-associated problems include:

- increasing cost of public transport
- lack of public transport or connections
- insufficient parking
- inappropriately parked vehicles
- a combination of private and commercial vehicles on the road.

Urban populations have boomed in developing countries as a result of population growth because people move to cities in search of work. The impact of this is intensified because these countries may not have the appropriate infrastructure to cope with the vast increase in population and, as a result, many transport systems are in a state of disrepair.

Housing

Another problem associated with population growth in cities, in both developing and developed countries, is insufficient and inadequate **housing.** Housing problems have come about as a result of social and demographic changes leading to an ever-increasing population, particularly in poorer countries. Other factors that add to housing shortages include:

- people are living longer
- people are choosing to marry later
- an increase in the number of single-parent families
- immigration from other countries
- rural to urban migration.

Land for development is in short supply in developed countries, particularly in cities, and affordable homes are difficult to find in urban areas.

Many mistakes were made in UK housing policy after the Second World War. The early approaches of re-housing people to unfamiliar areas (for example, New Towns Act) and demolishing homes to completely rebuild areas have been widely criticised. These policies meant that **urban renewal** was a process in which buildings were demolished and then rebuilt. This often meant that communities were split up and re-housed in high-rise buildings with little sense of community. There is now a drive to improve the infrastructure of existing areas and to provide more social and economic opportunities. This kind of **urban regeneration** means that people can continue to live in their own homes while work takes place around them.

The lack of available housing in **developing** countries, combined with rapid **urbanisation,** means that there are many housing problems. Many people who live in the city or who move there in search of work cannot afford housing. As a direct result, their only alternative is to build makeshift, temporary settlements or **shanty towns**. These are known as *favelas* in Brazil and *bustees* in India and have similar features, including:

- made from scrap material that is cheap and easy to find, such as wood and metal sheeting
- often without sanitation, water or electricity, resulting in widespread disease
- often overcrowded and uncomfortable.

THINGS TO DO AND THINK ABOUT

Think about the problems associated with urban areas in a developed and a developing country; how do these problems vary between the two and what impact do they have?

VIDEO LINK

Learn more about the problem of traffic in Britain by following the link at www.brightredbooks.net

DON'T FORGET

You are required to focus on housing and transport issues in both developed and developing countries. You should be able to comment on how these areas have changed over time.

ONLINE TEST

Want to test your knowledge of urban change? Head to www.brightredbooks.net

CASE STUDY – DEVELOPED CITY: GLASGOW 1

FUNCTIONS OF PRESENT DAY GLASGOW

Some more of Glasgow's present day settlement functions are as follows:

- **Industrial** – including factories and mining areas.
- **Educational** – Glasgow is home to Scotland's second oldest university and many other educational institutions.
- **Medical** – the Royal Infirmary, Southern General, Stobhill, Western, Victoria and Yorkhill hospitals. There is a concern that when the new South Glasgow University Hospital opens in 2015 the Victoria may close and the other hospitals may also be affected.
- **Entertainment** – UCG cinema, music venues, for example, the ABC and Bellahouston Park.
- **Transport** – Glasgow Central and Glasgow Queen Street train stations, Buchanan Street bus station, the underground system and Glasgow Airport.
- **Tourism** – home of the 2014 Commonwealth Games.
- **Residential** – areas of housing include Springbarn, Drumchapel and Knightswood.
- **Retail** – Sauchiehall Street, Buchanan Street and Argyle Street – including the Buchanan Galleries, Princes Square, Argyle Arcade and the St Enoch Centre.
- **Sport** – Sir Chris Hoy Velodrome (built for the 2014 Commonwealth Games, part of the sporting legacy) and football – Hampden, Ibrox and Celtic Park stadiums.

Glasgow is very well known for its retail functions. This includes high-end shops in the city centre as well as large out-of-town shopping centres such as Braehead and Silverburn.

ZONES/ZONING

Zone 1: commercial

Buchanan Street, Sauchiehall Street and Argyle Street are the backbone and main thoroughfare of Glasgow's shopping districts and up-market offices within the CBD and, in combination, make up the so-called 'Golden Z'. This pedestrian area is a paradise for consumption as it includes the covered shopping centres of the Buchanan Galleries, Princes Square, Argyle Arcade and the St Enoch Centre, as well as the shops on the main thoroughfares. Glasgow is the largest retail area outside London and much of its success can be attributed to pedestrianisation. Consumers are attracted to the unique and colourful Barras weekend street market, the smaller, characterful outlets of the city's bohemian West End, and a wide range of high-class cafés, restaurants, pubs and wine bars. The regeneration of these public spaces has been extremely influential in Glasgow's prosperity.

Buchanan Street, Glasgow – a busy, pedestrianised shopping street.

contd

Recent changes in the CBD

- Pedestrian safety in areas of Sauchiehall Street, Buchanan Street and Argyle Street has been dramatically improved as through traffic has been reduced. This has made the area much more enjoyable for shoppers.

- Altered road networks have been adapted to include one-way systems and bus/taxi/cycle lanes, which have resulted in smoother traffic flows through and around the city and encouraged commuters to use the existing public transport systems.

- Modernised city centre shops/shopping malls, such as the St Enoch Centre and Buchanan Galleries, have increased the appeal of the CBD shopping centre as retailers have worked hard to ensure consumers have the best possible all-round experience, while also countering the attraction of out-of-town shopping centres.

- New state-of-the-art offices, apartments and hotels have reinstated the appeal and increased the status of the CBD for businesses. The Merchant City is an outstanding example of how this can be achieved.

- Renovated tourist facilities and cultural attractions have allowed the development of successful short- and long-stay city breaks for tourists.

- Multi-storey car parks have made access to the city attractive and easy with their key locations, for example, next to the Buchanan Galleries, making shopping more convenient.

ONLINE

Learn more about Glasgow and other urban models at www.brightredbooks.net

DON'T FORGET

You should be aware of recent changes in different zones in both a developed and a developing world city.

Zone 2: residential

Glasgow faced a major housing shortage after the Second World War and adopted a policy of replacing run-down tenements with high-rise flats. The slum areas of tenements were demolished and high-rise flats were built in areas such as the Gorbals. The initial high expectations of these developments soon diminished as it was realised that they lacked any real sense of community and they deteriorated into dingy, ill-kept dwellings. Many communities had been dispersed without much consideration and the high-rise flats increased social exclusion and other social issues such as alcoholism and depression. Areas such as Easterhouse and the overspill New Towns like East Kilbride did not succeed as planned.

Regeneration strategy: GEAR project

The Glasgow Eastern Area Renewal (GEAR) Project was initially set up in 1976 to tackle the problems of economic decline, while trying to rectify and learn from some of Glasgow City Council's previous housing errors. Shortly before this time, there had been plant closures and the loss of many highly skilled and semi-skilled jobs in the East End of Glasgow. The remaining East End population was disproportionately made up from elderly and disabled people, those on low incomes and people with ill health and high mortality rates.

The changes brought about by GEAR ensured that, by 1986, two-thirds of the population were living in modernised housing. GEAR also created over 2000 additional jobs for Glasgow residents between 1976 and 1985. By 1987, the Scottish Development Agency had made available 190 ha of industrial land, which was planned to attract industrial investment into the area to boost the local economy. However, despite these successes, a devastating 16 000 jobs were lost between 1976 and 1985 as a result of the economic downturn. This not only damaged these efforts, but also substantially lowered morale.

ONLINE

Follow the link at www.brightredbooks.net for tasks on this case study.

ONLINE TEST

Revise your knowledge of this topic by taking the test online at www.brightredbooks.net

THINGS TO DO AND THINK ABOUT

The urban topic is a great area of the course on which to base your assignment. If you can study the comparative aspects of a city, this is a really good talking point for your findings.

CASE STUDY – DEVELOPED CITY: GLASGOW 2

ONLINE

Learn more about the legacy of the Glasgow 2014 Commonwealth Games at www.brightredbooks.net

ZONES/ZONING

Recent housing changes: 2014 Commonwealth Games

The 2014 Commonwealth Games brought new life into Glasgow – in particular, they injected a new vibrancy into Glasgow's housing areas. The Commonwealth Games Athletes' Village in the East End of Glasgow will be developed as both social housing and houses for sale. The aim is to improve and regenerate this area beyond the Games.

The village provided accommodation for 6500 competitors and officials during the Games. After the Games, the 38.5 ha site at Dalmarnock in the East End of Glasgow was transformed into a new housing development aimed at providing a sustainable and fresh approach to community living, with amenities and public space. The development included 300 houses for private sale, 100 for mid-market rent and a further 300 affordable homes for rent through Glasgow Housing Association, Thenue Housing Association and the West of Scotland Housing Association. This innovative project has also provided skills development, training and job opportunities. The roads and infrastructure in the area have also been improved.

Glasgow wanted to succeed where many other host countries had not, and wanted the Games to leave a positive legacy long after the crowds had dispersed. Glasgow has learned from the mistakes of other governments – for example, Greece spent over £7 billion building or upgrading venues for the 2004 Olympic Games and Beijing, which hosted the 2008 Games, struggled to find tenants and to raise finance. For Glasgow, the sporting success was hugely important, but the ongoing legacy was paramount in the approach to the Games and the changes implemented in the housing and transport infrastructure.

Glasgow's Athletes' Village

The Athletes' Village has been used to re-home Glasgow locals whose homes were demolished as part of the regeneration of the city. The homes are highly energy-efficient, which helps to cut energy bills, and all have wheelchair access. In addition, the 60 000 items of furniture used by the 6500 athletes were distributed to tenants who required them.

The success of the village has set a new benchmark for other sporting events. Not only has Glasgow provided a range of affordable housing for its residents, but it has created a sense of cohesion and legacy in its new communities. Several programmes contributed to this success, including:

- a Summer of Sport, run in partnership with Glasgow Life, for older tenants of Glasgow Housing Association, Cube and Loretto is keeping the over-60s fit and active as they take part in indoor bowls, golf, curling, mini-javelin and Jenga competitions

- a Gold Medal Programme supported 30 school projects and involved 10 000 pupils in Glasgow Housing Association and Cube communities – delivered in partnership with Glasgow City Council, the programme inspired pupils to try a new sport and encouraged an active lifestyle

- more than 900 pupils from 26 primary schools in Glasgow's East End took part in the FARE Mini Commonwealth Games, which was supported by Glasgow Housing Association and was held at the Emirates Arena.

contd

Some of the advantages of the new Athletes' Village include:

- 1440 houses (apartments, terraced, semi-detached and detached houses)
- over 1100 homes available to buy, 300 provided as social rented housing and a 120-bed care home for the elderly
- the homes are energy-efficient, helping to cut household bills
- local residents benefit from new services in the local area, including primary schools, shops and a community centre (the Dalmarnock Legacy Hub)
- improved sports facilities, including the Sir Chris Hoy Velodrome, which encourage health and well-being in Glasgow.

The disadvantages of the Athletes' Village include:

- not all of the original residents can afford to live there
- some residents reported that the money offered in compensation was insufficient to cover the cost of the move.

Glasgow City Council's top ten legacy statements of how the Games positively affected the city's people, schools, businesses and environment are as follows:

ONLINE

Visit the link at www.brightredbooks.net to find out more about the regeneration of the River Clyde.

TOP 10	
	12,000+ pupils from 160 schools participated in Healthy World learning project
	£198m invested in new/improved sports facilities since 2009
	97,000³ of contaminated soil treated and reused at Athletes' Village
	£145m of conferences and events secured due to Host City status
	£200m of tier 1 Games-related contracts won by Glasgow companies
	3,000+ Commonwealth Apprentices employed to date
	14,800 attendances at coaching courses since 2009
	400 bikes for hire at 31 locations through Mass Cycle Hire Scheme
	88% Glaswegians believe Games are positively impacting city
	1,500 Host City Volunteers welcoming visitors to sports and cultural venues

Glasgow City Council's top ten legacy statements.

LEGACY 2014
XX COMMONWEALTH GAMES
GLASGOW

Glasgow's vision also included these vision principles:

- a prosperous Glasgow
- an active Glasgow
- an international Glasgow
- a greener Glasgow
- an accessible Glasgow
- an inclusive Glasgow

Glasgow's promise to put 'community' at the heart of the Games is evident. Councillor Gordon Matheson said 'Glasgow 2014 will leave a lasting legacy and it is my ambition that it will be a People Legacy'. Glasgow achieved this by extensive consultation, including the Glasgow Household Survey and Have Your Say questionnaires and workshops. Throughout the process, the people of Glasgow were made to feel included and their opinions valued. It was realised by both the city's residents and the organising bodies that the Games had the potential to improve health, reduce inequalities and help to achieve a more sustainable Glasgow.

 DON'T FORGET

Some of Glasgow's changes have come about as a result of the 2014 Commonwealth Games; however, many of them have been underway for years.

 THINGS TO DO AND THINK ABOUT

How successful do you think the changes to housing (in relation to the Commonwealth Games) have been? Make sure you are aware of a range of factors that lead to urban change.

ONLINE TEST

Test yourself on this topic online at www.brightredbooks.net

CASE STUDY – DEVELOPED CITY: GLASGOW 3

ZONES/ZONING

Zone 3: industrial

ONLINE

Learn more about
these developments
by following the link at
www.brightredbooks.net

Glasgow is famous for its industrial heritage. Glasgow saw its first commercial steamship launch (of the *Comet*) on the River Clyde in 1812. It was this pivotal moment in the city's history that secured it as a centre for shipbuilding and industrial excellence. By the 1940s, the River Clyde was teeming with cargo ships and pleasure boats. The river had found its purpose and, like today, people were central to the river's activity and success. The Clyde was the powerhouse of Scotland and was the birthplace of many famous ships. Today, the Clyde is still influential in the city's success and is used for retail, leisure and educational purposes, including the Glasgow Science Centre; the Riverside Museum; The Tall Ship; Titan Clydebank; the SECC; Braehead; and Springfield Quay.

The River Clyde in Glasgow.

Examples such as the **Titan Crane** epitomise the transition from the Clyde's undeniably successful industrial past into its modern day triumph as an educational hub and showcase for the river's past and present achievements. The crane helped the world-famous John Brown's Shipyards to build some of the biggest and most famous ships in the world, including the *Queen Mary*, the Royal Yacht *Britannia* and the *QE2*.

contd

Greater Govan and Glasgow Harbour housing changes

Housing regeneration is taking place all over Glasgow. Projects are planned, underway and completed in: the City Centre; the Pacific Quay; **Greater Govan and Glasgow Harbour**; Renfrew Riverside and Scotstoun; Clydebank and Erskine; and Old Kilpatrick and Dumbarton.

Glasgow Harbour

There is an area of 53 hectares of mixed-use, high-quality development in the west end of Glasgow's Clydeside Waterfront within the Greater Govan and Glasgow Harbour area. This phased project is one of Glasgow's **flagship projects** of regeneration on **brownfield** sites previously used for industry. The development of this land will lead to economic, tourist and social benefits for both Glasgow and Scotland as a whole.

The project is split into clear phases. Phase 1 was completed between 2005 and 2007; phase 2 is partially complete and further phases will follow. Plans to expand the project to contain **retail** and **leisure** facilities are now underway for Glasgow Harbour. The project was started in 2005 and should be completed by 2015, costing **£1.2 billion**.

Phase 1

- located at the former Meadowside Granary site
- award-winning residential district
- built by Bryant Homes, CALA City and Park Lane
- 650 apartments, duplexes and penthouses, all with either a waterfront or parkland view
- properties also feature high-quality specifications, basement parking, landscaped areas and concierge facilities
- completed in 2007

Phase 2

- the development will be completed in three stages
- five imposing 16–22 storey towers facing the River Clyde and an eight-storey linear block to the north
- 819 apartments, 95% with views of the river
- basement parking
- ground floor commercial zone at the western edge of the site
- integration of both private and public space within and surrounding the development
- stages 1 (282 units) and 2 (205 units) of Phase 2 have been completed
- stage 3 on hold

Phase 2 of the project has created much controversy as it has been compared with some of Glasgow's other high-rise buildings, particularly those built in the 1960s. The new developments have also been compared to the **Red Row flats** on the city's north side, which were famous for crime and anti-social behaviour in the 1970s. The true success of Glasgow Harbour's regeneration cannot be truly measured until the project is complete.

 THINGS TO DO AND THINK ABOUT

Think about how Glasgow's function has developed from its origins as a market town to the present day and how it is continuing to evolve.

 **DON'T FORGET**

Industry is crucial to Glasgow's success. The quote 'The Clyde made Glasgow and Glasgow made the Clyde' is particularly interesting with regard to Glasgow's industrial past.

 ONLINE TEST

Head to www.brightredbooks.net and test yourself on this topic.

CASE STUDY – DEVELOPED CITY: GLASGOW 4

TRANSPORT PROBLEMS

Transport problem	Reason	Facts and figures
Increasing numbers of cars	There are more cars on the road today	Glasgow has the largest volume of traffic in Scotland, peaking in 2008 Number of licensed motorised vehicles in Scotland doubled from 1.3 million in 1975 to 2.7 million in 2012
Commuters	Commuters converge on the city centre every day during the morning and evening rush hours	41% of Glaswegians commute to work by car (as driver or passenger), 30% by public transport and 27% walk or cycle
Few bridging points	Glasgow only has a limited number of bridges across the River Clyde; vehicles are funnelled into bottlenecks	Kingston Bridge is the busiest section of road in Europe, with 170 000 cars using its ten lanes each day
Narrow streets	The Victorian grid-iron streets in Glasgow's CBD were built before cars – the streets are narrow and this is made worse by parked cars	The grid-iron streets were laid out in the 1780s by property developers benefiting from increased prosperity

Negative consequences of increased traffic

Vibrations from traffic can damage buildings

Increased journey times

Increased traffic noise

More accidents

Increased pollution from vehicle exhausts

Road rage

Negative effects of increased traffic.

ONLINE TEST

Go to www.brightredbooks.net and test yourself on this topic.

Solutions to Glasgow's traffic management problems

Changes to Glasgow's transport system	Need for change	Benefits
Clyde Arc or 'squinty bridge'	Bottlenecks of traffic built up at the existing bridges – the Clyde Arc has eased traffic flow entering the city as it is further downstream	Reduced traffic flow and accessibility
Broomielaw Tradeston pedestrian bridge (£7 million)	Council planners hope this pedestrian crossing, linking Broomielaw and Tradeston, will breathe new life into one of the city's most run-down areas	Ideally positioned as a catalyst for investment, promotes confidence and has a fundamental role in regeneration of the river
Fastlink	Runs along the side of the Clyde and uses the city centre's main bus corridors to improve journey times and options	Connects Queen Street and Glasgow Central rail stations and Buchanan Street bus station Passes via the financial services district in Broomielaw to the SECC, across the Clyde Arc bridge to the Digital Media Quarter, on to Govan and then the New South Glasgow Hospital In future, it will also go to the Braehead Shopping Centre and Renfrew

contd

Changes to Glasgow's transport system	Need for change	Benefits
M74–M8 interchange	Controversial scheme to extend the M74 into and through Glasgow was announced in 2003 at a cost of £500 million (final cost £692 million) – the 8 km route links the M74 at Carmyle with the M8 south-west of the Kingston Bridge Many believe the M74 interchange will support regeneration of the East End and was integral in the delivery of the 2014 Commonwealth Games The road has caused a divisive debate about the benefits to Glasgow, particularly as half of all households do not have a car and the extension will add to noise and air pollution It directly contradicts the Scottish Government's aim to cut and manage carbon emissions, although significant benefits are clear	M74 extension was completed on time and under budget on 28 June 2011 Improved access to Glasgow from all directions Reduced traffic flows in the city centre by almost 21 000 vehicles per day Reductions of up to 5500 vehicles per day (30%) on local roads Road traffic has reduced by 23% on sections of the M8 between Baillieston and Charing Cross and it has taken around 11 500 vehicles per day (7%) off the M8 over the Kingston Bridge
Improved information and signage – Get Ready Glasgow for 2014 Commonwealth Games – £4.5 million invested	The Games Route Network ran from Scotstoun in the west to Tollcross in the east and was crucial to the success the Games	Changes included upgrades to CCTV and traffic signals, investment in more variable message signs, and improvements to junctions, signing and road markings
Subway modernisation	The third oldest underground railway in the world The Subway urgently needed modernisation SPT, supported by the Scottish government, invested £300 million and work is now underway	Refurbished stations New driverless trains Tunnel improvements
Electrification of rail links in the Glasgow area	50 km of railway will be electrified as part of the Scottish Government's investment in the line between Cumbernauld and Glasgow	The £80m electrification of the Cumbernauld to Glasgow Queen Street line is making good progress Electric services are now running, improving the overall quality and frequency of services in this area
Network of charging points in and around Glasgow for electric vehicles	Charging units are available at the Hydro, Scotstoun Sports Campus and Tollcross International Swimming Centre Plans to install charging units at the Glasgow National Hockey Centre, the Athletes' Village and Holyrood Sports Centre near Hampden	Visual signal of Glasgow's commitment to the environment Green image may attract tourism Glasgow companies can add a new dynamic to corporate branding
Smart card ticketing	Paper tickets (and associated waste) will no longer be used on the Subway	Speeds up the use of trains, buses and the Subway, increases connectivity This has reduced waste and is more sustainable
Queen Street station facelift	In February 2014 Network Rail launched their public consultation on the proposed £104 million transformation of Glasgow Queen Street station Edinburgh–Glasgow Improvement Programme is working with Network Rail and Buchanan Partnership to refurbish Queen Street Station and expand the Buchanan Galleries	The Consultation has had; Economic, social and environmental benefits Faster, longer and more reliable trains, more sustainable Rail is the lowest carbon mode of mass transport and is essential to a low carbon economy Faster and more frequent rail links
Dalmarnock Smart Bridge – cost £4.8 million, of which £1.9 million was a grant from the European Regional Development Fund	Dalmarnock station on the north bank was identified as a vital transport hub for visitors to the Games; many key locations are close by	The bridge has halved travel times from Dalmarnock station to the planned National Business District in Shawfield

ONLINE

THINGS TO DO AND THINK ABOUT

With reference to a developed world city you have studied, explain the impact of recent housing changes that have taken place in the inner city.

Head to www.brightredbooks.net to see an answer for this question.

CASE STUDY – DEVELOPING CITY: RIO DE JANEIRO 1

Location of Rio de Janeiro in Brazil

LOCATION

Rio de Janeiro is located on the east coast of Brazil and is the country's second largest city. It includes both sandy beaches and mountains. Rio de Janeiro is a **city in transition** in terms of development and is known as a global city as it is significant in the global economy, particularly as it will host the 2016 Olympic Games.

RIO DE JANEIRO'S NEED FOR URBAN MANAGEMENT

Rio de Janeiro will host the 2016 Olympic Games and the 2016 Summer Paralympic Games. This is the first time a Portuguese-speaking South American nation has hosted the Games and only the third time the Games have been hosted in the Southern Hemisphere. Both **tourism** and **industry** are two of the main sources of income for the city.

Artist's impression of the Olympic Games Park.

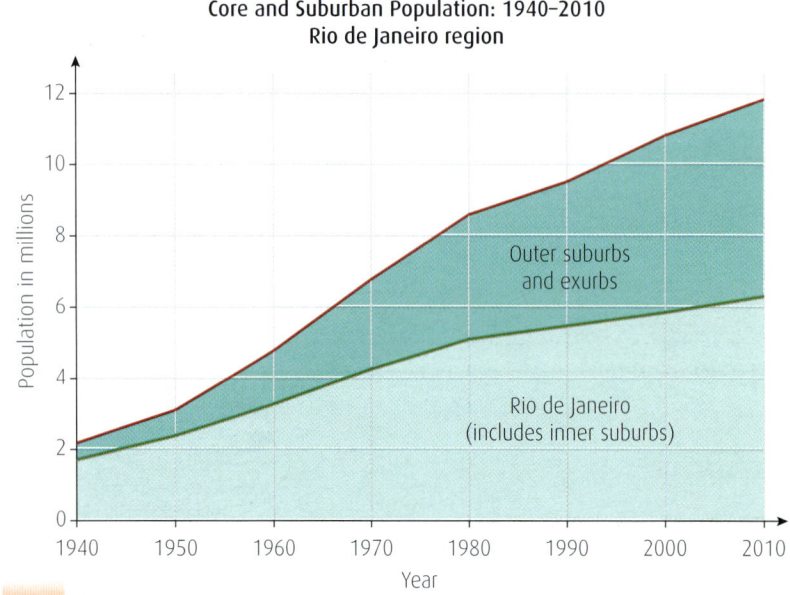

Core and Suburban Population: 1940–2010
Rio de Janeiro region

Outer suburbs and exurbs

Rio de Janeiro (includes inner suburbs)

Population in millions

Year

Rio de Janeiro's estimated population is six million and still growing. The combined pressure of the Games, tourism and an increasing population has created a need for urban management within the city.

Population of Rio de Janeiro from 1940 to 2010.

EXISTING TRAFFIC MANAGEMENT

Although the mountains surrounding Rio are very picturesque, they exacerbate the traffic flow problems by hemming in the city. A large amount of the city's traffic is forced along just a few main roads. This causes much congestion and air and noise pollution. Rio is famous for its traffic jams and poor urban traffic infrastructure.

The Metro and rail systems have been upgraded to try to improve the poor transport situation. Most journeys are made by private car. However, with an increasing population and increasing car ownership, this is unsustainable. As a result of its popularity with tourists and its metropolitan character, Rio de Janeiro was the most congested city in South America. Many changes were made to overhaul the transport system before the 2014 FIFA World Cup and the 2016 Olympic Games, including:

Traffic jam in Rio de Janeiro.

- improved and expanded subway lines

- improved and expanded train lines

- improved road infrastructure.

A recent study showed that the average drive through the city took 50% longer at peak times than at off-peak times – the number of cars has increased by 40% in the last 10 years. This study highlighted the work required to ensure the smooth running of large sporting events. To alleviate this traffic-related pressure, three extra public holidays were declared during the 2014 FIFA World Cup.

Two hot-spots for traffic congestion in Rio are the Niteroi Bridge and the Yellow Line Expressway – these both help to alleviate traffic congestion and are key areas where congestion builds up.

Rio's Niteroi Bridge

This bridge is a 14 km long and was one of the longest in the world when it was opened in 1974. The bridge is a shorter alternative to the 80 km road journey or one-hour ferry crossing connecting Niterio and Guanabarra Bay. However, it is enclosed and rising levels of car ownership have led to congestion, adding to Rio's traffic management problems.

Yellow Line Expressway

The Yellow Line Expressway is 21 km long and connects Barra da Tijuca in the west of Rio to the north and the international airport. Part of the expressway consists of two 2 km-long tunnels under the mountains of Tijuca National Park. The expressway was opened in 1997 and has a toll bridge to meet the huge costs of its construction. Approximately 70 000 vehicles use the road every day – this exceeds the authorities expectations by over 13 000. However, the expense was justified and economically viable as the expressway has reduced traffic congestion by around 40% on surrounding local roads.

 VIDEO LINK

Learn more about urban management in Rio de Janeiro by watching the clip at www.brightredbooks.net

Rio's Niteroi Bridge.

 DON'T FORGET

Rio is called a 'city of contrast' for many reasons. Can you identify why this is the case and make a list of reasons?

 ONLINE TEST

Want to revise your knowledge of this case study? Head to www.brightredbooks.net and take the topic test.

 ## THINGS TO DO AND THINK ABOUT

You can talk about the Rio de Janeiro FIFA 2014 World Cup and the impact this had on the city. What ongoing impact do you think the 2016 Olympic Games will have on the city's housing and transport systems in the future?

CASE STUDY – DEVELOPING CITY: RIO DE JANEIRO 2

RIO DE JANEIRO'S CHANGING TRANSPORT SYSTEMS

Metro Rio

The Metro Rio was founded in 1979 and runs over 41 km with 35 stations and several commuter rail lines. It is owned by the state and mainly serves the city's working class. The Metro has significantly reduced traffic congestion in the city, as well as increasing accessibility.

Future plans include a third line to Barra da Tijuca and Zona Oeste, as well as the expansion of Line 1. An additional line, Line 4, is planned for completion by 2016.

The Metro is Rio's safest and cleanest form of public transport. Subway cars can be overcrowded, but the Metro is a cheap and fast way to commute for tourists and locals alike.

Trams

Rio has the oldest operating electric tram system in South America. The system is now mostly used by tourists. The Santa Teresa Tram (*bondinho*) has been preserved both as a piece of history and as a quick, fun and cheap way of moving around the city for tourists. Many commuters now use the Bus Rapid Transit (BRT) system to cross the city.

Airports

Rio de Janeiro has two airports: Galeão International Airport and Santos Dumont. An investment plan for Rio's international airport was unveiled in 2009 ahead of the 2014 FIFA World Cup and 2016 Olympic Games. The plans included:

- renovation of Passenger Terminal 1 at a cost of US$314.9 million – most of this work was completed by May 2014
- completion and renovation of Passenger Terminal 2 at a cost of US$284 million
- construction of further parking at a cost of US$220 million
- 26 new boarding bridges and 68 check-in counters at Passenger Terminal 2 by April 2016
- modernisation and expansion of the aircraft apron
- up-dated infrastructure, including new restaurants and shops
- improved signage
- modernised elevators.

Much of the ongoing work at the International Airport is in preparation for the 2016 Olympic Games and includes large-scale renovation of the two main terminals. In total, US$2 billion dollars will be invested between August 2014 and April 2016 and passenger numbers are expected to increase to 60 million per year. It is claimed that the renovation will rival that of Changi International Airport, Singapore.

Bus lines

The main form of public transport in Rio is the bus service. There are 440 municipal bus lines, serving over four million passengers each day, and buses run more frequently during peak times. However, this is not a large number of buses compared with other world cities. Public transport in Brazil and Rio has been the target of many critics and was the catalyst for the **2013 protests**, which started in São Paulo and spread throughout the country. People were dissatisfied with the low quality and frequency of service together with fare increases. However, the buses are reasonably cheap, although buses with air conditioning cost more than those without. Concerns have been raised by both staff and passengers about travelling safely at night.

Improvements to the bus service have been made in preparation for the 2016 Olympic Games. In 2011 the creation of exclusive bus lanes or BRT systems with streamlined bus-stops improved traffic flow, particularly in Copacabanca, Leblon and Ipanema, as well as the city centre. The first of four BRT corridors, Transoeste, was opened in June 2012 and

Rio de Janeiro tram.

ONLINE TEST

Test yourself on this case study at www.brightredbooks.net

DON'T FORGET

The Brazilian Government is investing nearly US$17 billion in development ahead of the 2016 Rio Games.

linked Santa Cruz with Da Tijuca via a 32 km corridor. This line has halved travel times between the two areas.

Another three BRT corridors are planned to open by 2016. The first corridor, Transcarioca, is already being built and will connect Barra da Tijuca, the heart of the Games, with Galeão International Airport. The second corridor, Transolímpica, will link Barra da Tijuca and Deodoro, two important zones for the 2016 Olympic and Paralympic Games. The third corridor, Transbrasil, will provide access from Avenida Brasil to the competitors' zones and access in and out of the city.

The Transolímpica BRT system has been developed specifically for the Olympic Games and is estimated to cost US$400 million. It will require the construction of a 23 km express line and includes 18 stations and two terminals. The operation and maintenance of this project will be subject to a 35-year concession. The system will operate with 86 articulated buses to serve about 95 000 passengers each day and includes the construction of a 1.53 km tunnel and 48 bridges and viaducts.

The Transolímpica project is classified as a legacy project and will benefit 400 000 people daily and reduce travel times in the city to 40 minutes from 1 hour 50 minutes. To ensure all four lines are implemented successfully, the city has received US$1.5 billion from the state government for the Transbrasil line.

Bicycles

Bicycles are increasingly encouraged in Rio and are frequently seen on city streets. Many bicycle lanes extend over vast areas and bicycle hire and bicycle stands are common. The Bike Rio scheme began in October 2012 and is sponsored by the municipal government in partnership with Banco Itaú. Six hundred bicycles are available at 60 rental stations in 14 neighbourhoods throughout the city.

With over 90 miles of bicycle paths, Rio's improvements with regard to cycling are really starting to have an impact. The bicycle paths take in the areas of Copacabana, Ipanema, Flamengo and Lelbon, as well as a path circling Lagoa. An additional 90 miles of path are planned.

Summary of Rio de Janeiro's public transport changes and impacts

Transport system changes	Need for change	Benefits
Metro Rio	The system is still small, with only two lines and 35 stations, although there are plans to add a third line It is always very crowded, particularly in the rush hour	Metro Line 1 is being extended, adding an extra six stations between Ipanema and Barra da Tijuca The 14 km extension will increase capacity by 230 000 passengers per day – scheduled to open in December 2015
Tram	Existing services poor and extremely busy Trams often carry more than their capacity and can de-rail as a result, causing deaths and destruction	Network will have 42 stops serving the business district, the Porto Maravilha area, the Olympic press village and Santos Dumont Airport New contract to build and operate a 28 km tram network in Rio de Janeiro 14 km section to open by the end of 2015 in time for 2016 Olympic Games
Airports	The airports are 'tired' and have a low capacity for such a popular tourist destination	Modernisation of facilities, for example, Passenger Terminal 1 Expansion with more parking and new infrastructure, e.g. Passenger Terminal 2 Improved signage Modernised lifts Modernisation and expansion of aircraft apron
Bus lines	Rio's most used form of public transport – low quality and infrequent services	Exclusive bus lanes increased connectivity, improved traffic flow and commuter times in the city More bus corridors to be opened ahead of 2016 Games
Bicycles	Encouragement of bicycle use has taken the strain off many other transport systems, which were overused and sometimes dangerous Rio needed to make itself more attractive to tourists ahead of the 2016 Olympic Games and to improve sustainability	Increased sustainability Visual signal of Rio's commitment to environmental values Green image may attract tourists 600 bicycles available at over 60 rental stations 90 miles of bicycle paths

 THINGS TO DO AND THINK ABOUT

Think about the changes made to Rio's transport system so far and what advantages and disadvantages they bring. Now answer the model question below.

Rio de Janeiro, like many developing world cities, suffers from severe traffic congestion.

For Rio de Janeiro, or any named city you have studied in the developing world, evaluate the strategies used to manage traffic congestion.

 ONLINE

Head to www.brightredbooks.net to see a model answer.

CASE STUDY – DEVELOPING CITY: RIO DE JANEIRO 3

RIO DE JANEIRO'S EXISTING HOUSING PROBLEMS

Housing problems are common in cities such as Rio de Janeiro because social and demographic changes are leading to a greater demand for housing. People are living longer as a result of better diets, better access to medical care and improved sanitation. People are also choosing to marry later and there has been an increase in the number of single-parent families. Immigration also increases demand for housing. Rio has become a more desirable place to live because of the benefits, including jobs and investment, associated with the 2016 Olympic Games.

However, building new, affordable homes in urban areas is challenging because land is in short supply, given Rio's mountains to the north and west and the sea to the south and east of the city. Rio suffers from severe overcrowding and a shortage of housing.

The housing shortage in developing countries often leads to the development of unplanned housing (squatter settlements or shanty towns). These settlements are often unsightly, result in sanitation and pollution problems, and tend to have high rates of crime.

These settlements often develop as a result of rural to urban migration and immigration. Areas such as Rio cannot cope with large influxes of people and the added pressures on housing and infrastructure; 65% of the urban growth within Rio is a result of migration. Migrants who cannot afford housing are forced to build temporary accommodation in spontaneous settlements. In Brazil, these shanty towns are called *favelas*.

In shanty towns, houses are:
- made from scrap materials such as wood and metal sheeting
- often without services such as sanitation, water or electricity
- usually very overcrowded.

DON'T FORGET

This links to the section on rural-urban migration, pp. 44–45.

RIO'S HOUSING ZONES

There are clear zones of various kinds of housing settlement within Rio. As might be expected, the high-class residential areas are next to the CBD. They extend along the beaches (for example, Copacabana Beach) and are home to the wealthy workers of the CBD as they are adjacent to high-profile offices.

Barra da Tijuca is a modern coastal suburb in Rio's West Zone, next to the Tijuca Lagoon. The exclusive apartments have 24-hour security and armed guards to offer both security and prestige. Many of the residents worry about crime (for example, violence, thefts, muggings, abductions and vandalism), although the risk is far lower than in other parts of the city where these problems are directly associated with high levels of unemployment and extreme poverty resulting from the increasing population. Areas such as the Barra da Tijuca have benefited from the 2016 Olympic Games, as the existing services have been further improved to include a self-contained American-style city. The area has everything that residents require: well-paid office jobs, hospitals, schools, a university, parks, schools, restaurants and leisure facilities such as cinemas. The area is well connected with the Yellow Line Expressway, monorails and numerous dual carriageways. Barra da Tijuca also has a 5 km corridor of shopping malls and hypermarkets, making it the largest retail complex in South America.

Middle-class residential areas have grown up near the airports and highways, which allow commuters easy access to the CBD. These areas also suffer from the effects of Rio's infamously high rates of crime.

The least economically affluent people live further inland from the CBD and the main transport networks. These are the areas with shanty towns.

SQUATTER CAMPS

Various squatter camps have sprung up within Rio on unused and often unsafe land. These camps are used by the hundreds of immigrants who come to Rio every day in search of a better standard of living. These people have often moved from rural areas where there is only limited and low-paid work. These migrant workers can often not afford the high house prices in Rio and instead build their own dwellings from tarpaulin, scrap wood and corrugated iron. These dwellings are often built illegally on land that is not owned by the migrants. As a result, the government does not provide any infrastructure such as electricity and clean water, or services such schools, with the aim of discouraging them from staying.

These areas have a high incidence of diseases such as cholera and dysentery and a low life expectancy. As they become established, squatter camps tend to grow into larger *favelas*.

ROÇHINA *FAVELA*

Favelas are usually located on the edges of the city. Roçhina is Rio's oldest *favela* and has a well-established and constantly growing population of over 150 000 people. This area is located on a hillside to the south of Rio and looks onto the attractive beaches of Copacabana and Ipanema, which many of the migrants view as a symbol of success. Over 1.2 million people live in these conditions in Rio, often on less than £1 per day. Infant mortality rates are high in the *favelas*, 50 per 1000 births compared with a national rate of 15 per 1000 births.

Roçhina was first established in the 1940s, but grew at a much faster rate in the 1970s and 1980s. Many migrants settled here during the building of Barra da Tijuca, attracted by the construction jobs in the area.

Slum housing on a hillside in the Roçhina *favela* in Rio.

Roçhina has better living standards than the squatter camps because it is so well-established and has been improved over time. The area has standpipes providing running water, unpaved roads and shared toilets.

Many residents use their dwelling to provide a service and to earn an income. However, although many improvements have been made, Rochina still has the problems of a typical *favela*, including crime, overcrowding, disease and extreme poverty.

Street dwellers in Rio de Janeiro.

 DON'T FORGET

Brazilians refer to Rio – the second largest city in Brazil – as the 'marvellous city' (*cidade maravilhosa* in Portuguese).

 THINGS TO DO AND THINK ABOUT

You should be able to transfer some of the information about rural to urban migration provided in the population unit to this topic. This background information will help you to add depth to your exam responses, giving the reasons why so many people migrate to the city from rural areas and what impact this has on both the countryside and the city.

ONLINE TEST

Head to www.brightredbooks.net and test yourself on this topic.

CASE STUDY – DEVELOPING CITY: RIO DE JANEIRO 4

IMPROVING RIO'S *FAVELAS*

The authorities in Rio initially wanted to bulldoze these unsightly illegal areas before the 2016 Olympic Games. However, it was soon realised that people would simply move to another location and begin again. Before the announcement of the 2016 Olympic Games, two main methods of improving *favelas* were used in Rio.

Brazil is announced as the host for the 2016 Olympic Games.

Self-help schemes

These small-scale projects use the skills of local people to upgrade their own housing and neighbourhood. It gives people a sense of ownership and community spirit. Ahead of the Games, Rio did not have the capital to build large-scale housing zones, but it did have enough to provide residents with basic materials to improve their own homes.

Much of the housing in Roçhina has been upgraded and most of the homes are now made of more durable materials, such as concrete and bricks. Most (75%) of the homes in this area now have electricity and many services have been provided, such as shops, cafes and banks.

Site and service schemes

The site and service schemes function slightly differently to the self-help schemes because the locals are not in control; the local authority steers the building projects. The **Favela Barrio** (Slum to Neighbourhood) project began in 1994 and was an attempt to bring *favelas* in line with the rest of Rio. This project targeted smaller *favelas* with around 500–2500 homes. Brick houses were built and sanitation and water supplies installed. The residents were given the right to purchase their own homes. The facilities of the area were also radically improved, including building health clinics, schools and sports areas, and streets were widened for emergency services and refuse collection.

Success or failure?

These two types of scheme have been successful and many people have benefited from the improvements. The installation of water pipes and sewers has improved health and reduced disease, improving facilities in the city as a whole. The Favela Barrio project also generates income for the council as families pay rent for the new homes, which allows the funding of further improvements. However, it does not help families without jobs who cannot afford to pay rent.

HOUSING CHANGES AHEAD OF THE GAMES

Brazil's government anticipates that Rio de Janeiro will benefit from the global attention, tourism, infrastructure and economy resulting from the 2016 Olympic Games. The Rio 2016 Organising Committee predicts that the city will have a range of employment opportunities and training ideal for *favela* residents who want to better themselves. A total of 4000 temporary and permanent employees will be hired, 48 000 people will receive professional and volunteer training and 60 000 people will have the opportunity to volunteer. The real testament to the Games will be how they approach the management of the *favela* areas and their impact on the city's appeal during this time.

Cities hosting the Olympic Games have an associated sense of national pride and often see great potential to elevate their nation's status in the world. Rio is no different and there is a strong desire to rid the city of its underlying urban and social problems.

However, the issue facing the people of Rio is far deeper than relocating those who live in the city's 1000 shanty towns. Rio's authorities have received global attention by taking a very heavy-handed approach to tackling the crime inherent within the *favelas*.

CRIME CLEAN-UP

Rio is infamous for its vicious drug cartels, such as the Red Command. These groups make their own rules and dispense their own justice within their communities through private armies in areas such as the Vila Cruzeiro, which is run by drug warlords.

This kind of crime was almost unchallenged by the authorities until a police helicopter was shot down by a gang in the Morro de Macaco *favela* in 2009. The government was forced to take action to recover control. They sent in Pacifying Police Units to take back and consolidate state control in these gang-controlled slum areas. About 40 of the 1000 *favelas* have been occupied by Pacifying Police Units.

One such renowned pacification intervention took place in November 2010. This included a five-day battle within the Villa Cruzeiro and Complexo do Alemao *favelas* and involved 3000 heavily armed police and military forces in an attempt to clean up crime, with a particular focus on drug trafficking. The success of this kind of clean-up divides opinion both in the *favelas* and globally. During this particular operation, 37 people were killed and 200 arrested, but many escaped to neighbouring *favelas*. The heavy-handed approach of armoured vehicles, machine guns and police dogs has been widely criticised. Many within the *favela* feel that this is encroaching on their freedom. There have been reports of the Pacifying Police Units implementing curfews, bans on motorcycle taxis, bribes and aggressive behaviour as control tactics.

However, this is balanced with reports of far fewer murders, fewer gunfights and robberies, rising property values and increased tourism in the slums. These examples have left opinion divided as to the approach and success of the Pacifying Police Units.

DON'T FORGET

Only seven in every ten people living in a *favela* have access to a toilet or sewer pipe to transport waste away from their home; this lack of sanitation will affect people's ability to work and provide for themselves as they may be more susceptible to disease and illness.

DON'T FORGET

You need to be able to refer to both a developing and developed area you have studied, as you may be asked about both in the exam.

ONLINE TEST

To take a test on Rio de Janeiro, head to www.brightredbooks.net

THINGS TO DO AND THINK ABOUT

You need to be able to recognise both the positive changes that have been made and the less successful changes. Has Rio de Janeiro's government succeeded in its crackdown on crime within the *favelas*? Make a list of the positive and negative impacts. Think about **tourism** first to get started.

In the exam you will need to cover the following for the Urban topic:
- the need for management of recent urban change (housing and transport) in a developed and in a developing world city
- the management strategies used
- the impact of the management strategies

IMPACT AND MANAGEMENT OF RURAL LAND DEGRADATION

CAUSES OF RURAL LAND DEGRADATION

Rural land degradation describes how one or more of the land resources (soil, water, vegetation, rocks, air, climate, relief) have changed for the worse. This can be as a result of human activities (for example, grazing animals) or may occur naturally (for example, by wind and/or water). With today's ever-increasing global population, rural land degradation is often caused or worsened by the production of food for our growing population. Rural land degradation can signify either a temporary or permanent decrease in the productivity potential of the land. Rural land degradation is categorised as reversible or irreversible, depending on the location and the amount of damage that has occurred.

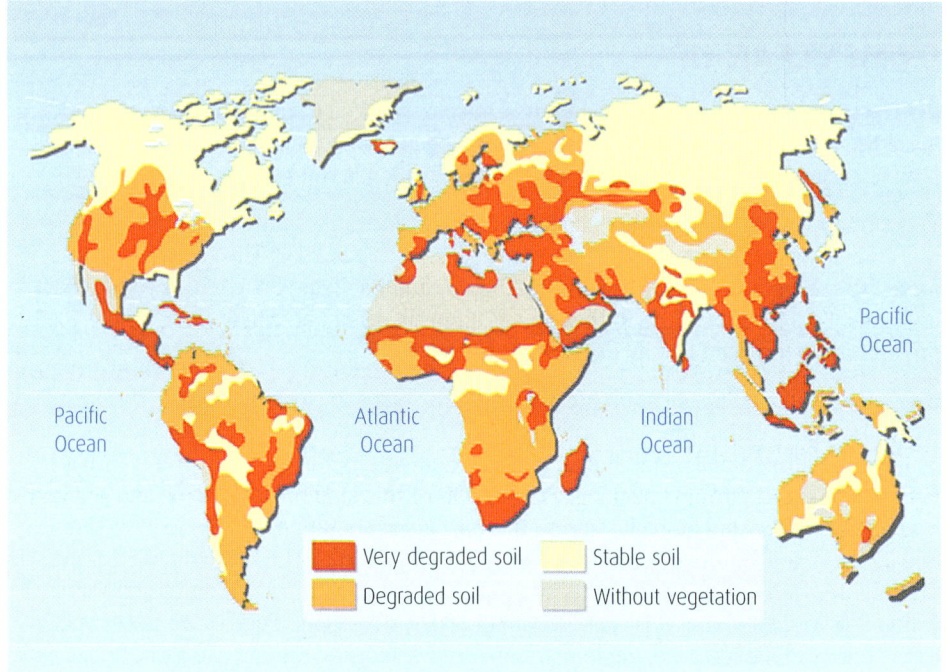

Soil degradation around the world.

Rural land degradation is difficult to assess as no single measure can be used, but it is imperative that it is managed as one-quarter of the world's population depend on drylands for their livelihood. Indicators are used to assess rural land degradation. The condition of the soil is one of the best indicators of land degradation. Soil performs many important processes, including supporting the growth of vegetation, the overland flow of water, infiltration, land use and land management. Soil degradation is, therefore, a key indicator of land degradation.

Soil can take hundreds of years to form and, like much of the Earth's surface, is susceptible to erosion. If soil is eroded faster than it is formed, then we say that the soil is becoming degraded.

Developed countries can afford expensive solutions to prevent and slow down land degradation, such as agricultural fertilisers, but developing countries struggle to pay for these.

contd

Effects of soil degradation.

Wind	Waterlogging	Decline in fertility	Water
Increased rock cover	**Influences on soil degradation**		Loss of vegetation
Sedimentation	Lowering of the water table		Increase in salts

Soil erosion by water

Soil erosion by water occurs in three stages:

- **detachment** – soil particles are detached from the main body of the soil by individual raindrops hitting the soil or by overland flow

- **transportation** – soil particles are carried downhill by floating, dragging or splashing

- **deposition** – soil particles are laid down, for example, on a river, lake or sea bed

There are four main types of soil erosion by water:

- **rain splash** – the impact of raindrops on the soil surface

- **sheet wash** – the removal of a thin, almost unseen, layer of surface soil

- **rill erosion** – the creation of very small eroded channels across a soil surface

- **gully erosion** – the creation of large gullies by large amounts of water flowing over the soil surface

Other factors influencing soil erosion

Steepness of slope – steeper slopes increase the speed of water flow and erosion.

Frequency and intensity of rainstorms – during violent summer thunderstorms, intense rain does not have time to soak into soils and flows over the surface.

Depth and permeability of soil – deep porous soils absorb more water.

Vegetation cover – plants protect soil from water erosion by slowing down and absorbing some of the water.

RURAL LAND DEGRADATION IN THE SAHEL REGION

Rural land degradation in acute semi-arid areas is called **desertification**. In the Sahel region, the main form of land degradation is soil degradation and the region is one of the most severely affected in the world. Desertification means land turning into desert. This is caused by a combination of physical and human factors. Land degradation in the Sahel region is thought to be caused by a combination of climate, drought, decreasing rainfall and human activity.

The name Sahel has been used for over 100 years and refers to the southern fringe of the Sahara Desert. The area stretches for approximately 5000 km from the Atlantic Ocean and there is no distinct border. Variations in rainfall throughout the region are reflected in the vegetation and varying ecosystems. The south, which receives more of the Tropical Maritime air mass, is denser in vegetation than the dry north.

THINGS TO DO AND THINK ABOUT

Think about the processes involved in the various types of soil erosion by water. Try to make a mind-map to highlight these with the aid of diagrams.

 ONLINE

Research the global effects of soil degradation further by following the link at www.brightredbooks.net

 **DON'T FORGET**

Land can be degraded through physical, biological and chemical action. Wind and water erosion are the main causes of land degradation. Vegetation binds soils and protects them from erosion, so when vegetation is removed, soils may be degraded.

 **DON'T FORGET**

Remember to revise specific examples and case studies of methods used by local people to try to combat the effects of desertification. These include agroforestry, water harvesting, magic stones, micro-dams/diguettes and terracing.

 **DON'T FORGET**

This links to the section on atmosphere, pp.36–37.

 ONLINE TEST

Want to revise your knowledge of the impact and management of rural land degradation? Head to www.brightredbooks.net

CAUSES OF RURAL LAND DEGRADATION IN THE SAHEL 1

Location of the Sahel region in Africa.

PHYSICAL CAUSES OF LAND DEGRADATION

The main physical cause of rural land degradation in the Sahel is the climate, which is directly related to its proximity to the Equator and the movement of the ITCZ. The ITCZ was covered in the Atmosphere section of this revision guide, see pp. 36–37.

The graph shows the variability of rainfall in this area.

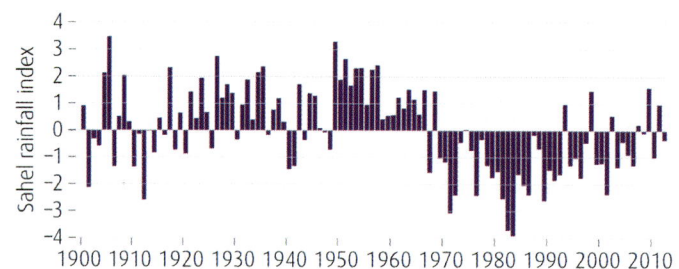

Climate graph for the Sahel region

HUMAN CAUSES OF LAND DEGRADATION

Increase in population

The human causes of rural land degradation are directly linked to the over-use of the land, including overgrazing, overcultivation, vegetation clearance and urbanisation. There is a direct correlation between increases in population and rural land degradation. An increase in population means that more demands are put on the land for resources such as food and fuel. This extra strain causes plants to fail, and the soil loses its structure and can easily be blown or washed away.

Overgrazing

The term **carrying capacity** refers to the maximum number of plants or animals that a piece of land can support without degradation.

Overgrazing quickly results in land degradation when too many animals are kept in one area. Repeated activity results in the removal of vegetation and the land is easily eroded when animals trample the ground. This is made worse when animals are kept in the same area for a long period and the ground has no chance to recover.

In times of drought (the dry season, October–April), people migrate south and move their animals around waterholes or wells, which can become overgrazed. Removal of the original vegetation means that the soil is exposed to wind and heavy rain. Animals trample and compact the soil, reducing infiltration rates and increasing surface run-off. This causes two problems: increased soil erosion by water and a reduction in the amount of water in the soil (for example, a lower water table), resulting in less water for future use.

Overcultivation

The economies of countries in the Sahel region rely mainly on agriculture production, with 80–90% of the population dependent on farming. Recurrent drought is common as a result of the variability and overall decrease in rainfall. Agriculture in the Sahel relies on rainfall and an intensive input of capital and labour.

Vegetation clearance

Deforestation occurs when land use is changed by clearing trees. Examples include ranches, roads and farms. This change in land use can have negative effects and lead directly to land degradation.

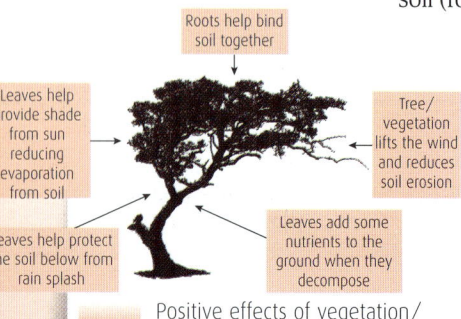

Watering hole in the Sahel.

Roots help bind soil together

Leaves help provide shade from sun reducing evaporation from soil

Tree/ vegetation lifts the wind and reduces soil erosion

Leaves help protect the soil below from rain splash

Leaves add some nutrients to the ground when they decompose

Positive effects of vegetation/ tree growth on soils.

contd

Urbanisation

Urbanisation leads to an increase in demand for firewood and, as a result, large areas of forest may be cut down and the wood turned into charcoal for easy transportation. This process is inefficient and half of the energy of the wood is lost during conversion.

CONSEQUENCES OF LAND DEGRADATION IN SEMI-ARID AREAS

Poverty

Desertification leads to social, economic and cultural consequences in the affected area. Poverty means that the limited natural resources are further exploited and abused. This, in turn, triggers a cycle of deprivation. Desertification affects the lives of many people living in semi-arid areas and the numbers affected are growing as the population increases.

Through this misfortune of global location and the lack of natural resources, many people have become vulnerable to global economic factors. These regions cannot support their population and often need to borrow from elsewhere. Drought and desertification lead to a reduction in the overall productivity of the area and the population has to turn to foreign products to survive. Although this is expensive, it is often more detrimental and costly to produce food locally than to import goods.

Migration

When rural populations can no longer survive in these harsh conditions, many migrate. Desertification can cause whole communities to disperse and local cultural traditions may be lost. The attraction of cities with jobs and housing does not always live up to expectations. Many migrants from rural settlements find themselves living in urban slums.

Dependence on external support

Rural land degradation can result in an over-dependency on external aid. Where the effects of land degradation and drought have been most acute, people have come to rely on food from other countries, for example, the Band Aid campaigns of 1985 and 2005.

Salinisation/waterlogging

Salinisation refers to the build-up of salts in soils. When farmers irrigate their crops, fast evaporation rates bring salts up through the soil profile to the surface, often preventing plant roots from accessing the water. Both rain water and water used for irrigation contain very low concentrations of salts. When water evaporates from a dry surface, a salty residue is left behind. Salinisation is most common on poorly drained soils as water rises to the surface rather than travelling down the soil profile.

Farmers frequently irrigate their crops in areas such as the Sahel. These areas are poorly drained and are rarely or never completely flushed through with water. As a result, salinisation is extremely common.

If poor farming methods or management result in waterlogging, then the land will eventually become bare as vegetation cannot grow successfully under these conditions.

Education and health care

Many of the countries in the Sahel require payment for health care and education. These have both suffered as a result of rural land degradation and associated poverty. This feeds back into the cycle of deprivation.

ONLINE

Investigate the humanitarian needs of the Sahel region by following the link at www.brightredbooks.net

ONLINE

Learn more about the Sahel region by following the link at www.brightredbooks.net

DON'T FORGET

Human activities such as deforestation, over-cultivation and overgrazing have contributed significantly to desertification.

THINGS TO DO AND THINK ABOUT

Try to think about the connection between the annual growth rate required for food production for a growing population and changes in cropland areas. How can this impact be better managed?

ONLINE TEST

Test your knowledge of the impact and management of rural land degradation at www.brightredbooks.net

CAUSES OF RURAL LAND DEGRADATION IN THE SAHEL 2

MANAGEMENT OF LAND DEGRADATION IN SEMI-ARID AREAS

Incorporating the three elements in the diagram below into farming strategies allows the **sustainable** management of rural land degradation where there is limited access to education, capital and technology. The schemes are entirely dependent on local people investing in the process and being able to continue independently if the technology requires maintenance or funds run out.

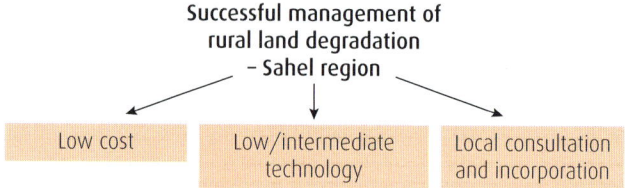

Successful management of
rural land degradation
– Sahel region

| Low cost | Low/intermediate technology | Local consultation and incorporation |

Management of rural land degradation in the Sahel region.

DON'T FORGET

Ecosystems in the African Sahel are adapted to varying degrees of aridity. The people who live there are very skilful in adapting their farming practices. The success of many of the strategies used relies on local people taking them forward.

SUCCESSFUL STRATEGIES IN THE SAHEL REGION

Irrigation

Sub-Saharan Africa has an ample water supply if it is used efficiently; currently, less than 2% of the total renewable water resources are used. Greater use of the region's water would sustainably boost the production of staple crops. Past attempts at irrigation schemes have failed as a result of high investment costs, poor planning, a lack of consultation with local people and a lack of ongoing support. Livelihoods in this area rely heavily on rainfall-fed agriculture. Irrigation allows farmers to move from subsistence to commercial farming. Schemes should be initiated by demand from farmers themselves; there is an increasing need for local management and maintenance of these schemes. Successful schemes have a participatory approach that encourages local farmers to take ownership of the scheme. Future proposals include transferring this ownership to land tenure to further **empower** farmers.

Irrigation failures	Key lessons
Irrigation schemes were too expensive	Participation, consultation and education allows farmers to take ownership of projects
Poor planning led to failure	Only technically sound initiatives work
Farmers were not given appropriate information and tools, and did not see the full benefits of irrigation	Governmental guidance needs to be clear and manageable
Expectations of improvements in yields were overambitious	Early land tenure decisions are essential in smallholder irrigation schemes
Infrastructure was not maintained and funding ran out	

Crescent-shaped terraces.

Terraces

Farmers have had to adapt to increasingly erratic rainfall patterns. Improvements in traditional farming systems are essential in increasing agricultural production. Farmers lack the knowledge, technology and tools to obtain the best from their fertile land, while competing with the rapid spread of desertification. Terraces are used, but most farmers build **traditional terraces** that do not make the most of the erratic rainfall. These terraces cannot withstand the strength of a monsoon storm, which can wash crops and soil away. **Crescent-shaped terraces** are better than traditional terraces and local farmers have been trained how to lay out and construct these.

contd

Afforestation/agroforestry

Planting new trees binds the soil together and provides shade to prevent moisture loss, a windbreak for crops and encourages nutrients to be returned to the soil. Depending on the species planted, trees may also provide nuts and fruit for humans and animals to eat. Tree planting needs to be sustainable and well managed if it is to be successful.

Stone lines

Stone contour lines allow rainwater to be used better and slow down erosion. They also increase water infiltration and capture sediment behind the semi-permeable barriers. Several organisations, such as Oxfam, have used this method very effectively and have incorporated traditional techniques from the Sahel into an improved system. In some instances, crop yields have increased by as much as 50%.

Tree planting in the Sahel.

Example

Burkina Faso experiences a four-month rainy season when downpours of Tropical Maritime rain are frequent; however, the rainfall is often destructive rather than beneficial to irrigation as the ground is dry and difficult to penetrate. A successful project in Burkino Faso, called Burkinabé-German PATECORE, ran between 1987 and 2006. This project introduced the simple, yet effective, solution of using stone lines, together with the more regular use of compost fertilisers. This adaptation has allowed the people of Burkina Faso to increase food production to feed the growing population.

Burkinabé-German PATECORE facts (1987–2006):

- **education** has played a key part – farmers know they must protect their existing agricultural land
- the project tackled more than 100 000 hectares of land
- 2.5 million cubic metres of stones have been stacked
- a subsidy of €250 per hectare has been given, a huge investment for West Africa
- during the dry season, the project helps 20 000 men and women to invest in their land – their most precious asset

Crops growing in Burkino Faso.

Changes to poverty and hunger	Changes to the environment
In the first year after implementation of the stone lines, agricultural yields increased by 50%	Tree growth in the area has increased by between 25 and 50%
Yields increased further with organic fertilisers – crops such as maize are now grown regularly	Contouring has increased the water table and has allowed vegetation to grow
Agricultural income has been secured	Intensification of agriculture on existing fertile land has decreased the need to convert drier areas into farmland – this would be both costly and labour intensive

ONLINE

Learn more about the Burkina Faso PATECORE project by following the link at www.brightredbooks.net

 ## THINGS TO DO AND THINK ABOUT

1 Discuss and note methods for preventing land degradation and rehabilitating degraded land in drought-prone areas, and suggest why the involvement of local people is important in this process.

2 Referring to either a named rainforest or semi-arid area, explain the techniques used to combat rural land degradation.

ONLINE

Find a model answer for the second question online at www.brightredbooks.net

ONLINE TEST

Head to www.brightredbooks.net to test yourself on the impact and management of rural land degradation.

In the final exam, you will be expected to know the following for the Rural topic:

- the impact and management of rural land degradation related to a rainforest or semi-arid area

MEASURING DEVELOPMENT 1

Inequalities **within** countries and variations in development **between** countries exist due to **physical, social** and **economic** reasons. Development and health are intrinsically linked.

DEVELOPMENT INDICATORS

The term 'development' indicates an improvement in the standard of living of individuals in any given country. Development levels are measured using development indicators between and within countries. Development indicators can be categorised into **social** and **economic** indicators. There is no single way to calculate the level of development because of the variety in economies, cultures and people. Geographers use a series of **development indicators** to compare the development of regions and countries.

ECONOMIC INDICATORS

Gross domestic product

The gross domestic product (GDP) is the economic value of all the goods and services produced by a country during the course of one year.

Vehicles per 1000 people

Cars and motor vehicles are expensive and they are a good indication of the wealth and development of a country.

Percentage of people employed in agriculture

Industry creates wealth and therefore rich, developed nations tend to have more people working in industry than in agriculture. A low number of people working in agriculture is an indication that a country is developed.

SOCIAL INDICATORS

Life expectancy

This is the average lifespan of someone born in that country. This can be affected by factors such as wars, natural disasters and disease.

Adult literacy

This is the percentage of the adult population able to read and write. In the UK, around 99% of adults can read and write to a fair level.

Infant mortality

This measures the number of children who die before they reach the age of 1 year old for every 1000 live births per year. The figure of 1000 is used so that countries of very different sizes can be compared.

COMPOSITE INDICATORS

Traditionally, only economic indicators were used to gauge a country's development. However, this did not give a sufficient overview and anomalies were often missed.

Composite indicators use a combination of indicators to give a better overview of development. Two main types of composite indicator are shown on page 75.

contd

Physical Quality of Life Index

The Physical Quality of Life Index (PQLI) uses three development indicators to make the result more reliable: life expectancy; infant mortality; and adult literacy.

Some countries may have, for example, a high literacy rate, but still be fairly undeveloped.

The rating is between 1 and 100. A value of 77 is considered satisfactory.

Human Development Index

The Human Development Index (HDI) uses the three development indicators of: health – life expectancy at birth; education – educational attainment; and wealth – GDP per capita, adjusted by taking into account the cost of living in different countries

The HDI has been used by the United Nations since 1990. The three indicators combine to give an average score between 1 and 0. The closer to 1, the more developed the country.

United Nations inequality-adjusted Human Development Index rankings for 2013.

HDI rank	Country	Human Development Index (2013)
1	Norway	0·944
14	United Kingdom	0·892
17	Japan	0·890
51	Bahamas	0·789
57	Russian Federation	0·778
88	Fiji	0·724
103	Mongolia	0·698
117	Phillipines	0·660
135	India	0·586
154	Yemen	0·500
186	Congo (Democratic Republic of the)	0·338

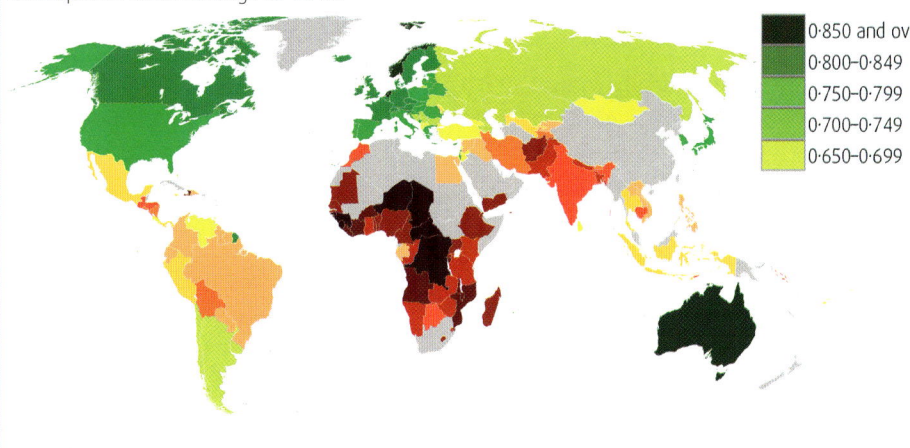

0·850 and over	0·600–0·649	0·350–0·399
0·800–0·849	0·550–0·599	0·300–0·349
0·750–0·799	0·500–0·549	0·250–0·399
0·700–0·749	0·450–0·499	0·200–0·249
0·650–0·699	0·400–0·449	Data unavailable

 ONLINE

Follow the link at www.brightredbooks.net for more on international indicators of human development.

ADVANTAGES OF COMPOSITE INDICATORS

- Averages disguise and distort internal variations – for example, a few people are immensely wealthy, but the majority are poverty stricken.

- A more balanced view of development combines education, health and the economy.

- Indicators may be inappropriate to the real quality of life in poorer countries, for example, number of televisions.

- Subsistence agriculture is not included in the wealth of a country.

- GNP may be inflated by oil revenues.

- Some regions are much better off than others, for example, urban/rural contrasts.

DON'T FORGET

Examples of composite indicators include the HDI and the PQLI.

 ## THINGS TO DO AND THINK ABOUT

It is very important that you fully understand the difference between economic, social and composite development indicators. You need to know how useful each individual indicator is at showing levels of development and be able to suggest reasons why a range of indicators is used.

 ONLINE TEST

Test your knowledge of measuring development online at www.brightredbooks.net

MEASURING DEVELOPMENT 2

PROBLEMS WITH DEVELOPMENT INDICATORS

The problem with development indicators is that they are an average figure for the whole country and often hide huge differences in the standard of living, inequalities and anomalies within a country.

National indicators – not a true reflection

There are several reasons why development is not accurately represented using national indicators. Reasons include the following:

- Huge **variations** in wealth between rich and poor, especially where the elite benefit from selling natural resources, such as oil in Saudi Arabia or diamonds in Zambia – in these instances the average income is artificially inflated by a small number of wealthy people.

- **Gender inequalities** between men and women – women have a lower standard of living than men in some countries – for instance, women have been denied education in countries such as Afghanistan under the rule of the Taliban.

- Different **ethnic and racial groups** can have a higher standard of living than the indigenous population – for example, in South Africa, despite the end of apartheid.

- Differences between **rural and urban areas** – urban areas have more access to services such as education, health care, schools and hospitals as they are concentrated in cities; rural areas may be remote due to the terrain, for example, the Amazon rainforest in Brazil or the Andes in Peru.

Malala Yousafzai, a campaigner for the education of women and girls.

- **Inaccurate information**, often caused by inherent difficulties in carrying out censuses and illiteracy – the GNP only takes into consideration goods and services for sale, which is inappropriate for rural areas because subsistence farming is not included.

DIFFERENCES IN LEVELS OF DEVELOPMENT BETWEEN COUNTRIES

Different countries in the developing world are at very different stages of development. There are several common factors that influence the levels of development both **within** and **between** countries. These include:

- **Climate** – areas with an extreme climate, for example, too hot, too cold, too wet or too dry, are usually less developed as food production is difficult.

- **Drought and famine** – many poorer countries are prone to drought and famine; some of the poorest countries, such as Mali and Chad, are in the Sahel region of Africa.

- **Extreme weather** – any extreme in the weather will make life difficult; it is difficult to build houses and roads, to farm, to attract industry and to earn a living if the weather is either very hot or very cold, for example, it is difficult to drill through the permafrost in the tundra regions.

contd

- **Natural hazards** – areas likely to be hit by floods, hurricanes, volcanic eruptions, earthquakes or drought tend to remain less developed.

- **Landscape** – some countries have terrain that is difficult to navigate or use, for example, mountains or areas of forest.

- **High population growth** – money is redirected to education and health for the young and cannot be spent elsewhere.

- **Endemic disease** – for example, malaria or AIDS, makes development difficult; productivity is decreased as many people cannot work and pay taxes; money is redirected to health instead of economic development and the health system often cannot cope with the large scale of disease.

- **Trade** – land-locked countries find it difficult to trade with other countries as they are more difficult to access.

- **Natural resources** – these can often increase wealth, for example, oil is in great demand and can be sold for a huge profit; however, factors such as war can prevent a country from developing.

- **Political corruption** – this hinders development as many investors avoid such a country and the economy will suffer.

- **Industrial development** – countries with industrial/manufacturing roots tend to increase in wealth, whereas countries that rely heavily on agriculture tend to remain less developed.

- **Civil war** – this leads to political instability and money is often spent on the military and weapons, not on people and health care; investment opportunities can also be lost.

ONLINE

Find out more about comparing levels of development between countries by following the link at www.brightredbooks.net

 **DON'T FORGET**

Individual development indicators can hide huge disparities.

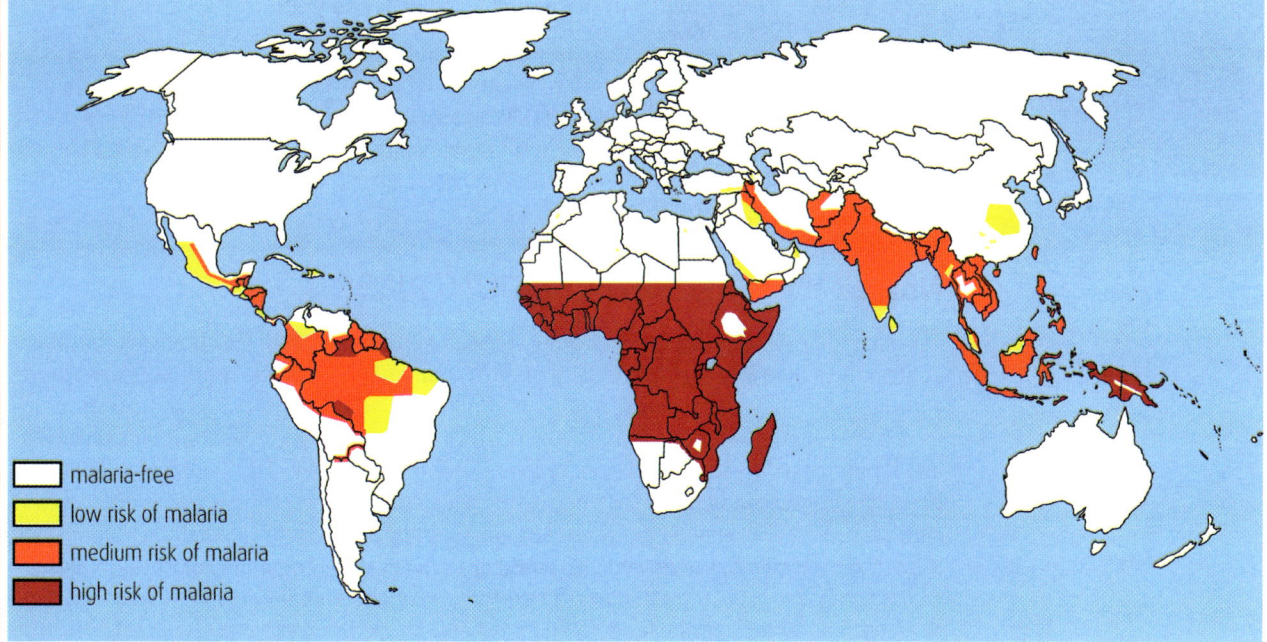

- malaria-free
- low risk of malaria
- medium risk of malaria
- high risk of malaria

Distribution of malaria.

 THINGS TO DO AND THINK ABOUT

Statistical information is not always what it appears to be at first glance. Huge variations between rich and poor countries can often be hidden by averages. Remember, there are advantages to using particular development indicators as well as disadvantages. Try to refer to both in your exam answer.

 ONLINE TEST

Test your knowledge of measuring development online at www.brightredbooks.net

VARIATIONS IN DEVELOPMENT BETWEEN COUNTRIES

We will now look at some of the variations in development between different countries in more detail.

Drilling for oil.

OIL

There are many reasons why there are wide variations in development between economically less developed countries. Some countries have natural resources that are in global demand. This enables them to trade with other countries and the money earned can be used to improve the living standards of their citizens. The United Arab Emirates and Saudi Arabia have large amounts of oil, which is in extremely high demand worldwide as a result of increased industrialisation and transport. As oil reserves are finite, it will eventually run out, making it a very valuable resource. However, countries such as Mexico, where oil reserves have already decreased by 75%, will soon have no oil to trade with and their economies will suffer as a result. Often, only a small number of people control the supply of oil, meaning that most citizens do not benefit from this increased ability to trade.

Countries such as Burkina Faso lack any resources with which to trade, hindering their development because they have very little money going into the economy. This lack of money means that health care is under developed and people in Burkina Faso have a life expectancy of only 56 years.

NEWLY INDUSTRIALISED COUNTRIES

Taiwan and South Korea are newly industrialised countries (NICs), which means that the contribution of industry to the GDP is greater than one-third of the total. NICs earn large amounts of money from the car manufacturing and electrical equipment industries and have a very high rate of economic growth. These countries have educated, skilled workers and labour costs are low because people are desperate to work. The capital brought in by selling goods can be invested into the infrastructure of the country, improving health care and education, which increases the living standards of the citizens. Cheap labour and imported materials have allowed shipbuilding to develop in Singapore's natural harbour, which has boosted both trade and the economy as a whole; unemployment rates have decreased as a result of these projects. Brazil and Mexico were among the first NICs, along with many South-East Asian countries such as Malaysia, Thailand and Japan.

SUBSISTENCE FARMING

Countries with a large proportion of people working in subsistence farming, such as Mali, do not generate large sums of money as they do not have much to trade and barely survive on what they do have. Long periods of drought also mean that growing enough crops is extremely difficult. Areas may suffer from famine and people may become ill and be unable to work. If the government intervenes to provide money and food, there will not be enough money to invest in education. Mali also has a lot of debt with other nations, so there is even less money available to improve the infrastructure as they try to settle these debts.

CIVIL WARS

Civil wars and political unrest have led to disruption in countries such as Somalia and Rwanda, which find it difficult to attract investors or trading partners. Corrupt governments mean that the people's needs are neglected as money is redistributed directly to government officials or dictators; this hinders the country's ability to develop as little money goes into infrastructure and services.

NATURAL DISASTERS

Natural disasters limit a country's ability to develop fully as money is spent on replacing the damaged infrastructure. Bangladesh experiences frequent floods and cyclones, and the damage can take decades to correct so that, just as the country has recovered from one natural disaster, another occurs. This deters investors and can seriously affect the popularity of the country for tourists.

The after-effects of natural disasters will limit development in countries such as Sumatra.

TOURISM

Tourism is very beneficial as it provides job opportunities and generates money that can be spent to improve areas with poverty and high unemployment, and to improve standards of living, such as in Jamaica and Thailand.

 THINGS TO DO AND THINK ABOUT

Using the headings given above, develop your own up-to-date case studies from online research. Put your findings into a mind-map. These are YOUR examples and will help you with your revision.

 ONLINE

Learn more about tourism in Thailand by following the link at www.brightredbooks.net

 ONLINE TEST

Test yourself on the variations in development between different countries online at www.brightredbooks.net

➕ **DON'T FORGET**

Development indicators are in some ways averages and can therefore hide real poverty – for example, the differences between the oil-rich families in Arab states and the poor majority population.

COUNTRIES AT DIFFERENT STAGES OF DEVELOPMENT

CASE STUDY – ETHIOPIA, NIGERIA, SAUDI ARABIA AND SOUTH KOREA

There are a number of reasons (both human and physical) that can explain the differences between Ethiopia, Nigeria, Saudi Arabia and South Korea. Both Saudi Arabia and South Korea were previously developing countries, but are now no longer classified in this way. The range of countries in the table below highlights the differences between developing countries and NICs. Saudi Arabia is now considered to be a very wealthy country as a result of its oil wealth, and the fact that many of its people live in poverty is disguised by its GDP ranking.

The first table compares a number of development indicators. Both Ethiopia and Nigeria are considered to be developing countries. However, Nigeria's GDP ranking suggests otherwise. Obvious differences between the four countries can be seen in the next table, which highlights important variations in terms of natural resources and other influencing factors.

Development indicators for Ethiopia, Nigeria, Saudi Arabia and South Korea.

Development indicator	Ethiopia (developing country)	Nigeria (developing country, except for GDP ranking)	Saudi Arabia (developed in terms of GDP and life expectancy)	South Korea (newly industrialised country)
GDP ranking	84	25	19	14
GNP per capita (US$)	410	1430	18030	22670
Mortality rate for children under 5 years old	68	124	9	4
Adult literacy (%)	39	51·1	82·2	97·9
Crude birth rate	33.5	41·5	19·9	9·6
Life expectancy (years)	63	52·1	75·3	81·3

Characteristics of Ethiopia, Nigeria, Saudi Arabia and South Korea.

Ethiopia	Nigeria	Saudi Arabia	South Korea
Population: 94.1 million	Population: >140 million	Population: 28·83 million	Population: 50·22 million
Limited natural resources	Oil resources	World's largest petroleum exporter; also produces and exports a variety of industrial goods all over the globe	Naturally occurring resources
Cycle of poverty	Tropical climate with lush vegetation and a diverse range of crops that grow all year	Strong export links (Asia, the Americas and Europe)	Strong export location and links (USA, Japan, China)
Land degradation and subsistence farming	Farming-related debt as farmers try to improve their land, but lack expertise and knowledge	Heavy subsidies, including taxation	Large, flexible, well-educated workforce
National debt	Traditionally large families mean more mouths to feed	Employment is healthy as the result of a successful public sector	Powerful allies (for example, the USA)

contd

Ethiopia	Nigeria	Saudi Arabia	South Korea
Civil war, corrupt government	High infant mortality resulting from poor access to education	Free health care and education	Secure government
Endemic diseases and poor health	Outbreaks of disease (for example, ebola) and poor health	Intermarriage between tribal groups accumulates existing wealth	Population growth under control
Low living standards as income in low	Low living standards as income in low	Wealth reinvested in housing, health, sanitation and food	Wealth reinvested in housing, health, sanitation and food
Lack of industrial development	Lack of industrial development	Reinvestment infrastructure, for example, new underground railway system for the capital to be built by 2018	Wealth based on industrial growth
		Wealth based on continued investment from multinational companies	Multinational companies keen to invest

The following diagram shows the ways in which we can identify whether a country is developed.

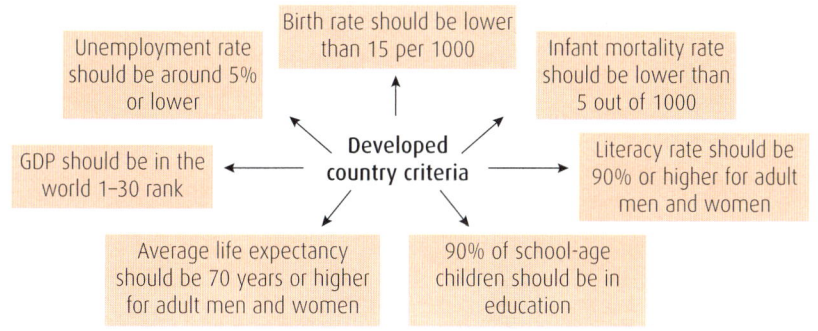

ONLINE

Revise this topic by following the link at www.brightredbooks.net

DON'T FORGET

You should refer to countries and statistics that you have studied in class in your exam answer when referring to varying levels of development between countries. Remember the examples in the case study.

DON'T FORGET

Read the question carefully, check the available marks and think about how to structure your response.

THINGS TO DO AND THINK ABOUT

- Using the criteria for developed countries in the diagram above, traffic light the table of countries for Ethiopia, Nigeria, Saudi Arabia and South Korea. This will help you to determine which of the countries are truly 'developing' and which are 'developed' in a sense. Remember, you can refer to countries that have passed through the initial stages of the demographic transition model and are now considered to be NICs.

- When answering a question about differences in development between countries, try to give both a range of examples and a good depth of answer with associated statistics, if possible. You should try to show a real understanding of why a country is at that stage in its development. Try answering the following question:

Q. Referring to named developing countries that you have studied, explain why there is such a wide range in levels of development between developing countries.

ONLINE

Head to www.brightredbooks.net to see a model answer.

ONLINE TEST

Test yourself on countries at different stages of development at www.brightredbooks.net

MANAGEMENT OF MALARIA

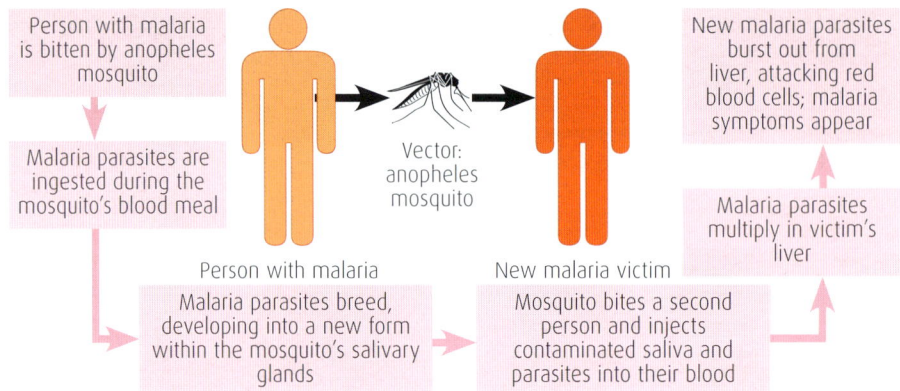

Cycle of transmission for malaria.

Person with malaria is bitten by anopheles mosquito

↓

Malaria parasites are ingested during the mosquito's blood meal

↓

Malaria parasites breed, developing into a new form within the mosquito's salivary glands

Vector: anopheles mosquito

Person with malaria

New malaria victim

Mosquito bites a second person and injects contaminated saliva and parasites into their blood

↑

Malaria parasites multiply in victim's liver

↑

New malaria parasites burst out from liver, attacking red blood cells; malaria symptoms appear

FACTS ABOUT MALARIA

- Malaria is caused by **parasites** that are **transmitted** to people through the **bites of infected mosquitoes**.
- **3.4 billion** people worldwide are at risk of contracting malaria.
- The disease is **endemic** – it is always present in the population.
- People in the **poorest countries** are most at risk.
- Between **500** and **600 million** people are infected by malaria at any one time.
- A child dies from malaria every **30 seconds.**
- Between **4000** and **8000** people die from malaria every day.
- Between **1.5** and **3 million** people die from malaria every year.
- Malaria is both preventable and curable.

One of the mosquito species that can transmit malaria to humans.

Malaria is found within the Tropics of Cancer and Capricorn where humidity is high and temperatures are between 15 and 40°C.

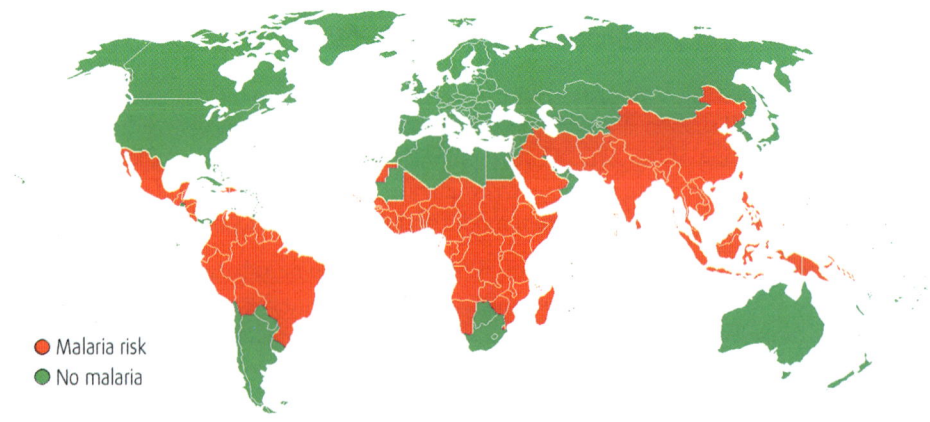

● Malaria risk
● No malaria

Areas of the world at risk from malaria.

Symptoms of malaria include fever, headache and vomiting. Symptoms usually appear between 10 and 15 days after a mosquito bite. If not treated, malaria can quickly become life-threatening by disrupting the blood supply to vital organs. In many parts of the world, parasites have developed resistance to the drugs used to treat malaria.

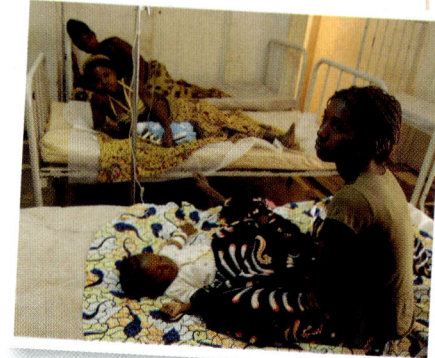

Babies and children can become very ill with malaria.

FACTORS AFFECTING THE TRANSMISSION OF MALARIA

The table shows the physical and human factors that allow malaria to spread.

Physical factors	Human factors
Humid, tropical areas with temperatures from 15 to 40°C – infection is spread under these ideal breeding conditions for mosquitoes	Lack of money – poor people cannot afford nets to protect themselves from mosquito bites
Isolated/remote areas where medical care is not available	Poor sewerage/irrigation/paddy fields/areas of stagnant water
Easily contracted	Houses provide shade areas for mosquitoes to digest meal of blood
Drought/arid conditions (mosquitoes can breed in hoof prints)	Government cannot afford drugs to treat malaria
	Mother-to-child transmission
Water supply affected/restricted/contaminated	Humans provide blood meals for mosquitoes
Increased levels of disease	Proboscis of female mosquito enters blood stream through skin and prevents clotting of blood
Vegetation – shade for mosquitoes to digest meal of blood	Poor nutrition – easier to contract disease if ill/malnourished as a result of low immunity
Rainfall levels	Stagnant water in paddy fields and outside sources of water provide areas for mosquitoes to lay their eggs
Female anopheles mosquito lay eggs in areas of stagnant water	
	Exposure of bare skin allows mosquitoes to bite

CONSEQUENCES OF MALARIA

Malaria affects the infant mortality rate as babies and children can become very ill. Adults may become too weak to work, which leads to a loss of productivity. Many people may not be able to provide enough food for their families and will remain poor and more susceptible to disease. The country's resources will be used for health care rather than improving services, infrastructure and education, which further hinders development and leads to a lower standard of living.

If malaria is controlled, the workforce will be more productive and the GNP will increase. If people are unable to work as a result of disease, tax receipts are lower and less money is collected to put back into the prevention and cure of malaria. Tourists are less likely to visit the country if they are worried about contracting a disease such as malaria, so controlling malaria may increase tourism – the money raised could be redirected to other types of health care or education. Marshy areas that are drained to remove mosquito breeding areas can be brought into agricultural production, providing more food for the local population or for export.

 THINGS TO DO AND THINK ABOUT

You should be able to refer to both the physical and human factors that lead to the spread of malaria. In particular, think about what how to minimise the spread of malaria by altering the human factors.

 ONLINE

Learn more about the effects of malaria by following the links at www.brightredbooks.net

DON'T FORGET

Both physical and human factors lead to the spread of malaria.

 ONLINE TEST

Go online to www.brightredbooks.net and test yourself on this topic.

METHODS USED TO CONTROL MALARIA

Malaria is an entirely preventable and treatable disease. In treating malaria, the primary objective is to ensure the rapid and complete elimination of the plasmodium parasite from the patient's blood. This will prevent uncomplicated malaria progressing to severe disease or even death. Chronic infection can lead to malaria-related anaemia.

Insect repellents and sprays used in the fight against malaria.

ONLINE

Check out the link at www.brightredbooks.net for information on developments in vaccinations against malaria.

PROTECTING PEOPLE FROM MALARIA

Several methods are currently used to prevent the spread of malaria and to protect people from this endemic disease. These methods include the eradication of mosquitoes, the prevention of mosquito bites and prophylactic (preventative) drugs.

Despite these efforts, malaria still occurs in areas where there is both a high human and a high mosquito population, which leads to a high transmission rate. However, unless the parasite is eliminated from the whole world (as it has been from North America, Europe and much of the Middle East), it is possible that it could re-establish through the frequent and extensive travel and migration resulting from globalisation.

Most protection methods against malaria focus on **prevention**.

TREATING THE VECTOR (MOSQUITOES)

Method	Effectiveness
Use of **insecticides** such as DDT and malathion to kill mosquitoes	**Expensive** Oil-based and may be **harmful** to the environment Insects quickly build up **resistance**, so treatment becomes ineffective over time
Spraying larvae with **egg white** to suffocate them	Difficult to see if all areas are completely covered – waste of a vital food source, which could be used to improve people's immune system
Draining breeding grounds	**Impractical** – impossible to drain all areas of water and requires a great deal of labour **Paddy fields** cannot be drained without removing a vital food source
Genetically engineered sterile male mosquitoes – the species would die out as could no longer breed	**Ethical concerns** – alters nature and wipes out a whole species of insect
Use of **BTI bacteria,** which kill mosquito larvae – these are grown in fermented coconuts that are then broken open and thrown into mosquito-infested ponds; the bacteria are eaten by mosquito larvae and destroy their stomach lining; BTI bacteria can also be grown commercially in industrialised countries	**Safeguards ponds** for up to 45 days – coconuts are cheap and plentiful
Eucalyptus trees can be planted to soak up excess water, which decreases the amount of stagnant water available for mosquitoes to lay their eggs	Insects also stick to the trees and eventually **die from starvation**
Larvae-eating fish, for example, guppies/muddy loach, can be introduced to paddy fields	Relatively **non-disruptive** to agricultural production
Draining water behind dams	**Defeats the purpose** of dams, which is to store water

contd

TREATING THE HOST (HUMAN)

Methods	Effectiveness
Insect repellents, for example, DEET	This can be costly and has to be regularly maintained
Prevention through **education programmes** that inform people how the disease is spread and how to avoid it, e.g. covering bare skin to protect it from mosquito bites	Cheap and therefore fairly effective in the poorest countries
Insecticide-treated **mosquito nets**	Fairly cheap, but a single mosquito can transmit malaria
Treatment with **anti-malarial drugs,** for example, quinine sulphate or chloroquine	People can build up **resistance** to these drugs, making them ineffective; some drugs have serious **side effects**

ORGANISATIONS FIGHTING MALARIA

There are a number of organisations, such as the **Bill and Melinda Gates Foundation**, that aim to publicise the problem of malaria and to generate money to help developing countries.

Organisations such as the **World Health Organisation (WHO)** have launched campaigns to eradicate malaria using insecticides and drugs. They also conduct research into finding ways to cure and prevent malaria.

Following extensive trials, a preventative vaccine called RTS,S is proving successful. It is hoped that this could be widely used in malaria-endemic countries. Read more about this at www.malariavaccine.org

The **Red Cross** provides emergency aid and medical care to many areas suffering from the effects of mosquitoes. They provide both **short-term aid**, which provides training and education in **primary health care** strategies and also **long-term aid**, which helps to improve overall health in malarial regions.

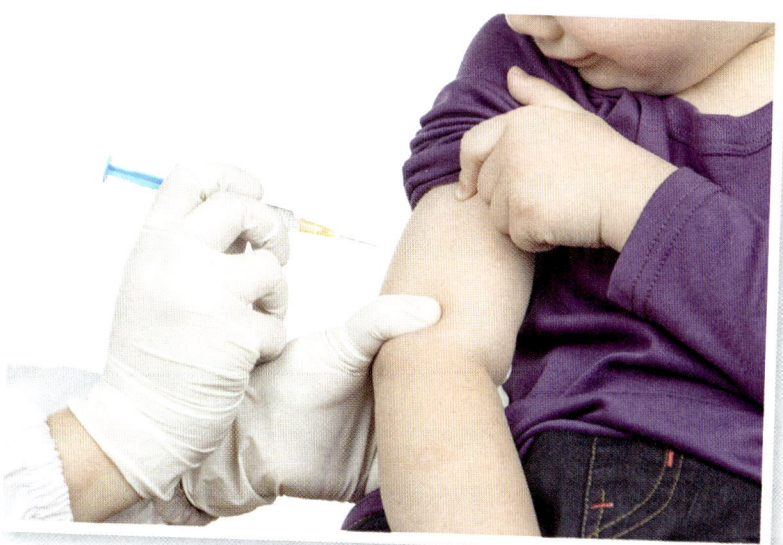

The Red Cross and Red Crescent have programmes to help combat malaria.

 ONLINE

Learn more about the Bill and Melinda Gates Foundation at www.brightredbooks.net

 **DON'T FORGET**

You should be prepared to comment on the strategies adopted to control the spread of malaria and their effectiveness. Can you comment on how expensive or how easily available these strategies are?

 DON'T FORGET

Malaria is a fascinating topic of study that allows you to produce interesting and up-to-date exam answers with relevant statistics.

 ONLINE TEST

Want to revise your knowledge of malaria? Head to www.brightredbooks.net and test yourself on this topic.

 ## THINGS TO DO AND THINK ABOUT

Learn your case study as this will help you to fully understand how malaria is being tackled and how successful this has been in a particular area. Some programmes have been more successful than others.

PRIMARY HEALTH CARE STRATEGIES

Developing countries face daily challenges in providing effective health care for their ever-increasing populations. These challenges are worsened by a lack of money, nomadic populations and remote regions. Many people in poorer countries live in rural areas, where roads are of a low standard and the terrain makes access difficult. To meet these challenges, primary health care is provided using local approaches.

DON'T FORGET

Remember the links between malaria-related primary health care strategies and more recent strategies reported in the news for the ebola epidemic.

PRIMARY HEALTH CARE IN DEVELOPING COUNTRIES

Primary health care strategies suit developing countries because local people are provided with a basic level of health care that is not too expensive and ensures a better quality of life.

Primary health care strategies may include:
- 'barefoot' doctors (BFDs)
- oral rehydration treatment
- vaccination programmes
- health education schemes
- local initiatives
- Water Aid – a charity that helps to provide safe drinking supplies.

ADVANTAGES OF PRIMARY HEALTH CARE STRATEGIES

Primary health care strategies bring many **advantages** including:
- improvements in health care and life expectancy
- provision of basic health care
- low training costs – trusted local people are trained to treat common illnesses, sometimes using cheaper traditional remedies
- communities thrive through self-help schemes
- healthier populations thrive
- mainly a preventative, rather than curative, approach
- BFDs refer more seriously ill patients to hospital, ensuring that no time is wasted
- overseas aid focusing on primary health care is more cost-effective than building a large hospital for a limited number of patients
- local labour and building materials are often cheaper – also provide training/transferable skills for locals and faster acceptance/usage in the community.

Example: 'Barefoot' doctors

- Local people trained in basic health care techniques and knowledge.
- Refer more seriously ill patients to hospital.
- Educate people about hygiene and sanitation.
- Educate people about family planning.
- Techniques include traditional medicine, e.g. acupuncture in China.
- Prescribe and distribute basic medicines.
- Immunisation programmes.

BFDs are preferred in many developing countries as these 'doctors' are trusted local people trained to recognise common diseases. They educate people about hygiene and sanitation, which is vital in ensuring that disease is not spread. Mothers are educated to care for their babies, reducing child mortality rates. Before BFDs, many people walked hundreds of miles to access medical care. Now they can visit a local 'doctor' who, if necessary, can refer them to hospital, avoiding wasting time, resources and money. BFDs were introduced in China in the 1960s and 1970s and were a huge success – China's one-child policy was reinforced through education and the availability of contraception. This method of health care is much cheaper than providing hospitals, which no longer have to deal with minor ailments that could be easily treated locally.

Mozambique

- BFDs were **unsuccessful** in the 1990s because of the civil war.
- Communications were difficult in the countryside as a result of land mines.
- Money was spent on weapons instead of health care.

VIDEO LINK

To learn more about the BFDs, watch the documentary at www.brightredbooks.net

contd

China

- BFDs here were **successful** between the 1960s and 1970s.
- People no longer had to travel large distances for health care.
- Two million BFDs were trained and many used traditional techniques such as acupuncture.
- BFDs were important in the one-child policy, providing education and contraception.
- Clinics for up to 1000 people promoted health, hygiene and treated minor ailments.
- Mobile health teams were able to train BFDs during times of low agricultural activity.
- Street community hospitals were used to treat up to 2500 people.
- BFDs were successful until 2003, but locals could then no longer afford health care – tuberculosis has since increased and immunisation rates have decreased.

The table lists some of the strengths and weaknesses of BFDs.

Strengths	Weaknesses
Improvement on previous system	Diagnoses were sometimes inaccurate
Approximately one doctor for every 40 families	Unable to provide enough health care
Fixed rate of payment, no over-charging	Medical staff lack skills (inappropriate prescriptions)
Introduction of health care to the countryside	Some teams unable to pay for further training, co-operative systems collapsed
Traditional techniques used – promoting local culture	People could not afford BFDs and migrated to urban areas
Less distance to travel to services	BFDs suffered from exhaustion – pressure of job and over work
	BFDs often illiterate – could not complete paperwork

PRIMARY HEALTH CARE STRATEGIES IN MORE DETAIL

Oral rehydration treatment

Every year, diarrhoea kills five million children under the age of five years old. The use of oral rehydration treatment is essential as it replaces lost bodily fluids. Unsafe water supplies and a lack of sanitation in many areas are often the cause of this problem.

Vaccination programmes

Vaccination programmes can be administered by BFDs. Although this is expensive, preventative treatment is less expensive than curative treatment. Six children die every minute as a result of inadequate immunisation. Each vaccine in the WHO's immunisation programme costs around US$3 per child. In 1974, the Expanded Programme of Immunisation aimed to immunise children against six common diseases, including polio and tuberculosis.

Provision of clean water – Water Aid

Primary health care also involves the provision of clean water supplies to prevent water-borne diseases such as cholera and malaria. Water Aid is a UK-based charity that provides support to local communities in developing countries. It aims to improve water quality by introducing sustainable initiatives for individual countries and local people that allow progress to continue when the charity leaves. Their work lowers levels of infant mortality and leads to a healthier workforce, who can then better support their families.

THINGS TO DO AND THINK ABOUT

Think about the appropriateness of the studied primary health care strategies.

In the exam, you will need to know the following for the Development and Health topics:

- validity of development indicators
- differences in levels of development between developing countries
- a water-related disease: causes, impact and management
- primary health care strategies

DON'T FORGET

Be prepared to comment on a range of primary health care strategies and their associated strengths and weaknesses.

ONLINE TEST

Take the test on primary health care strategies online at www.brightredbooks.net

CLIMATE CHANGE AND GLOBAL WARMING

CAUSES OF CLIMATE CHANGE

The climate of the Earth has changed throughout geological time. Scientists have found evidence for these changes by studying ice cores, rocks and fossils.

Air molecules trapped in ice cores can be analysed and can show evidence of changes in the Earth's atmosphere over time. Ice cores taken from Antarctica can be used to identify these subtle changes.

Scientists can also identify changes in rocks and fossils. They can compare current ecosystems with fossil ecosystems from the past, which will give information about temperatures, precipitation and the plants/animals of particular periods of geological time.

EFFECTS OF CLIMATE CHANGE

There is more evidence of climate change every year. Scientists have been studying regions that are at risk of drought, falling crop yields and flooding.

There is also substantial evidence of:
- reductions in **ice cover** – some scientists have suggested that the Arctic ice cover may not exist in future summers
- **glaciers** are retreating – less snow is falling and higher temperatures lead to melting
- **biodiversity** is changing – plants and animals are adapted to particular ecological conditions, for example, as the sea ice in the Arctic reduces, the hunting grounds of polar bears are becoming smaller
- **sea-level changes** – sea levels are rising as glaciers and land ice melt and the water they contain is added to the world's oceans – the Earth's albedo is also reduced as the ice cover recedes and more solar radiation is absorbed, exacerbating the rise in temperature
- **changing weather patterns** – some areas may receive less snow and rain; extreme weather events, for example, tornadoes, are occurring more frequently
- **global warming** – there has been a steady increase in global temperatures since the 1950s.

ONLINE

Explore this topic further online by following the link at www.brightredbooks.net

GLOBAL WARMING

The Earth is heated by energy from the Sun. Greenhouse gases in the atmosphere prevent this heat from escaping – without this natural 'greenhouse effect' the Earth would be about 40°C colder.

Greenhouse gases include:
- water vapour
- carbon dioxide
- methane
- nitrous oxide
- ozone
- chlorofluorocarbons (CFCs).

These gases all occur naturally, but greenhouses gases are also emitted via the activity of humans. The main anthropogenic sources of greenhouse gases are:
- burning fossil fuels (coal, oil and gas)
- felling forests (in tropical regions and elsewhere), which means there are fewer trees to absorb carbon dioxide
- animal waste, resulting in the emission of methane.

CFCs are important greenhouse gases. In addition to contributing to global warming, they can also damage the ozone layer in the Earth's atmosphere. The ozone layer protects the surface of the Earth from harmful ultraviolet rays that can cause sunburn and damage to plants.

contd

Temperatures have been steadily increasing since the Industrial Revolution. Your grandparents or great-grandparents may have memories of colder winters and warmer summers. Even rises in temperature of only 1°C can cause a shift in weather patterns around the world. Within the last four years, the UK has seen more stormy weather, longer periods of rain and more out-of-season rain.

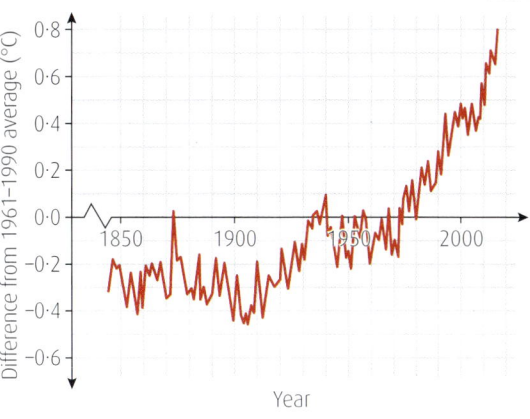

Changes in average global temperatures from 1850 to the present day.

 **DON'T FORGET**

Summarise your notes for the exam using mind-maps. Try to cover both the natural and human causes of global warning.

NATURAL CAUSES OF GLOBAL WARMING

Natural causes of global warming include:
- changing oceanic circulation caused by the melting of glaciers and land-based ice – this affects the El Niño and La Niña oceanic currents
- the release of large amounts of methane from melting permafrost
- volcanic eruptions – large amounts of volcanic dust in the atmosphere shield the Earth from incoming solar radiation, lowering global temperatures, for example, the eruption of Mount Pinatubo in 1991 caused a dip in global temperatures
- Milankovitch cycles – variations in the tilt and/or orbit of the Earth around the Sun – alter the amount of sunlight that the Earth receives and where this sunlight falls on the Earth's surface – on time scales of thousands of years, these changes are enough to start or end an ice age
- increased sunspot activity, which may raise global temperatures.

 ONLINE

Follow the link at www.brightredbooks.net to see the effects of volcanoes on global warming.

Melting glaciers

Dust from volcanic eruptions may lower global temperatures by reducing the amount of sunlight reaching the surface.

HUMAN CAUSES OF GLOBAL WARMING

Human causes of global warming include:
- burning fossil fuels – produces carbon dioxide
- car exhausts and nitrogen fertilisers – produce nitrous oxide
- methane – emitted from rice fields, landfill sites and farm animals
- increased industrialisation leading to air pollution – including nitrous oxides
- traffic fumes – produce carbon monoxide
- deforestation – increases amount of carbon dioxide in the atmosphere
- emission of CFCs from fridges, air conditioning units, aerosols and as by-products of producing polystyrene packaging (for example, pizza and burger boxes).

Methane from landfill sites contributes to greenhouse gas emissions.

Deforestation results in fewer trees to absorb carbon dioxide.

 THINGS TO DO AND THINK ABOUT

Using the graph of global temperature increase:
- describe the changes in temperatures since 1900
- explain why these changes have taken place.

 ONLINE TEST

Revise your knowledge of climate change and global warming by testing yourself at www.brightredbooks.net

MANAGEMENT STRATEGIES AND THEIR LIMITATIONS

SUSTAINABLE ENERGY

Sustainability in terms of energy means the management of resources or projects/industries so that future generations can also benefit. There is an increasing demand for energy to be sustainable, to both minimise damage to the environment and to minimise global warming.

It is important that:

- we develop new types of energy that are more sustainable than fossil fuels
- afforestation is encouraged rather than deforestation
- energy is used efficiently both by industry and in the home
- developed countries try to change their energy sources from fossil fuels to alternatives such as solar, wind, tidal and hydro-electric power
- developing countries are encouraged to be more energy-aware.

International concern about global warming and sustainable energy has led to many countries making a public commitment to reducing their carbon emissions.

DON'T FORGET

Other cities are doing their bit to become sustainable, for example Glasgow's MACH (Mass Automated Cycle Hire) scheme. This links to the section on Glasgow's transport problems, pp. 58-59.

CARBON CREDITS AND CARBON OFF-SETTING

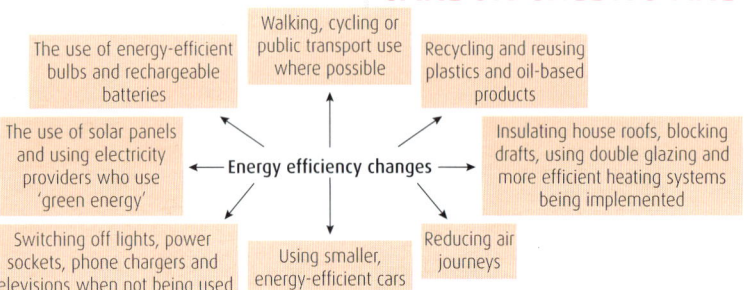

The use of energy-efficient bulbs and rechargeable batteries

Walking, cycling or public transport use where possible

Recycling and reusing plastics and oil-based products

The use of solar panels and using electricity providers who use 'green energy'

← Energy efficiency changes →

Insulating house roofs, blocking drafts, using double glazing and more efficient heating systems being implemented

Switching off lights, power sockets, phone chargers and televisions when not being used

Using smaller, energy-efficient cars

Reducing air journeys

Ways people can improve their energy efficiency.

Carbon credits are part of a scheme that aims to reduce greenhouse gas emissions. It is largely based on the idea of making polluters pay according to how much pollution they generate. If people pollute less, then they will pay less in carbon taxes. This scheme is sometimes known as '**carbon off-setting**'.

The diagram to the left highlights some of the ways in which people can make energy-efficient changes.

Case Study – London Congestion Charge

An example of a successful attempt to reduce both congestion and carbon emissions is the **London Congestion Charge**. Drivers are penalised financially for driving in the 'congestion' zone in London. The project aims to discourage people from entering into this busy zone during peak times and to maximise their use of public transport, resulting in reduced congestion, less time spent queuing, reduced pollution and a reduction in the cost of congestion to the economy. The money raised is reinvested into London's public transport system. Buses have been upgraded and older buses have been removed from service.

The successes of the London Congestion Charge include:

- a reduction in traffic congestion
- fewer accidents
- a reduction in air pollution
- more reliable and frequent bus services
- increased retail revenue inside the congestion zone
- increased investment in public transport.

A London bus.

HOW CAN WE BE MORE ENERGY EFFICIENT?

As the world's population continues to rise, we all have a responsibility to use energy more efficiently. Fossil fuels will eventually run out, so we have to find alternatives and we can all make a difference by becoming more energy efficient, which will also reduce our fuel bills.

We can all save energy by:

- insulating our homes to reduce heat loss, for example, insulating lofts, pipes and tanks – 'smart' metres can be used to monitor the efficient use of electricity and double glazing will reduce heat loss through windows

contd

- using energy-efficient bulbs
- switching off and unplugging appliances such as TVs and chargers when not in use
- using public transport and more fuel-efficient or electric cars
- recycling more waste, for example, plastics, glass and food waste.

EXAMPLES OF CLEANER ENERGY

We need to reduce our reliance on fossil fuels. More of our electricity is now being generated from renewable sources such as wind, solar, hydro-electric power and biomass.

If we reduce our use of cars, we will substantially reduce our emission of greenhouse gases. Many modern cars have 'stop–start' technology and electric cars are becoming increasingly popular. We can use public transport and encourage car pooling to reduce emissions of greenhouse gases.

We can also reduce greenhouse gases by making more conscious decisions about the products we buy. Buying local products rather than imported goods reduces the amount of fossil fuels used in transport. As consumers, we can decide to buy goods with less packaging.

Conventional farming uses chemicals that contain nitrous oxides and methane. Organic farming does not use these chemicals and therefore has less impact on global climate change.

THE ENERGY DEBATE

Most governments have promised to reduce greenhouse gases by signing up to the Kyoto Protocol, but the USA, one of the world's biggest greenhouse gas emitters, has repeatedly refused to agree to this treaty. There is also some concern that poorer countries would struggle to develop if they had to keep their levels of carbon emissions low. This calls into question the commitment of developed countries to manage climate change.

Some campaigners have objected to alternative methods of energy generation. There have been many debates about the construction of wind farms and their visual impacts. They are noisy and may affect local birdlife. Some campaigns have reached the national news, for example, Donald Trump and his objection to an offshore wind farm adjacent to his golf resort in Aberdeenshire.

The construction of hydro-electric dams involves flooding large areas – often areas of farmland or regions of historical/cultural significance.

THINGS TO DO AND THINK ABOUT

1 (a) Explain the human activities which have contributed to the changes in global air temperatures.

(b) Discuss the possible impacts of global warming throughout the world.

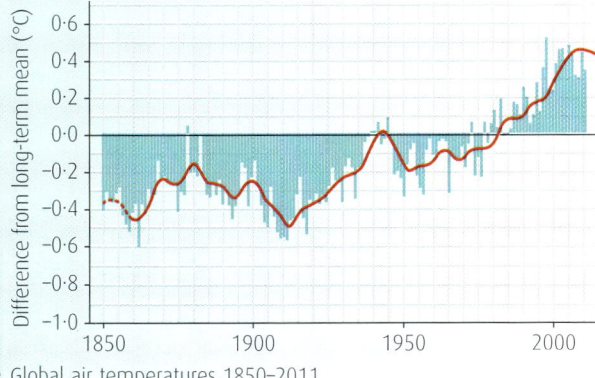

Global air temperatures 1850–2011.

Greenhouse gases, emissions by type.

Nitrous Oxide 4%
PFCs and CFCs 2%
Methane 10%
Carbon Dioxide 80%

 **VIDEO LINK**

The clip at www.brightredbooks.net presents the views on global warming of Al Gore, who was US Vice-President in the Clinton Administration from 1993 to 2001. Some scientists feel that this is an exaggerated view and that there is no need to manage climate change.

 DON'T FORGET

There are a number of ways to revise for the exam – taking/summarising notes, watching educational clips and practising questions from past papers.

ONLINE TEST

Test yourself on management strategies for global warming and their limitations at www.brightredbooks.net

 ONLINE

Head to www.brightredbooks.net to see possible answers to this question.

LOCAL AND GLOBAL EFFECTS OF CLIMATE CHANGE 1

EFFECTS OF CLIMATE CHANGE

Climate change causes much controversy between scientists, politicians and industrialists. However, regardless of the causes of climate change, global warming is real and there is a mutual consensus that the planet is warming. It is believed that global temperatures have risen by between 0.4 and 0.8°C over the past 100 years and are having a direct impact on sea levels and the weather. The United Nations has recently released a scientific report stating that climate change is happening now and will continue to happen for centuries to come. The report concluded that there is a 90% certainty that human activity is directly responsible for climate change and the report calls for a global approach to combat climate change. It is agreed that the effects of climate change are extremely serious. These effects can be categorised as environmental, social and economic.

ONLINE

Learn more about the effects of global warming by following the link at www.brightredbooks.net

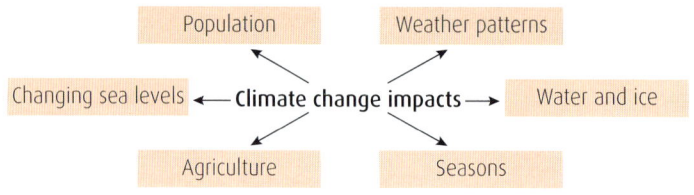

Effects of global climate change.

NEGATIVE EFFECTS OF CLIMATE CHANGE

Areas of impact	Effects	Negative consequences
Health	Increase in diseases such as malaria and cholera	More money needed to fight disease
Vegetation	Shifting of flora and fauna to different areas – extinction of some species	Spread of pests and disease and decline in crop yields may increase food shortages
Weather	More extreme climates in inland locations; more frequent and devastating natural disasters, for example, hurricanes	Financial and social costs from natural disasters will affect development
Oceans	Increases in sea temperatures, rise in sea levels, shifts in ocean currents	Changes in fish stocks and their location will affect fishing industry
Landscape	Reduced snow cover in some areas and melting glaciers	Rise in sea levels
Hydrology	Reduction in area of wetlands as precipitation is reduced – river flooding increases in other areas	Pressure on water supplies, creating problems for hydro-electric power schemes and an increased need for irrigation
Population	Reduction in areas of land suitable for human habitation, for example, coastal areas prone to flooding and low-lying areas of Bangladesh	Increased density of population, increased disease and malnutrition
Climate	Location of the jet stream is changing and areas of low atmospheric pressure are beginning to move further south and becoming more intense	Better forecasting is required to warn of approaching storms

ENVIRONMENTAL EFFECTS OF CLIMATE CHANGE

The whole Earth is warming, affecting global ecosystems. This means that many species are threatened with extinction.

Rising temperatures

Rising sea temperatures result in an increased risk of flooding.

contd

Forests

Forests and woodlands provide natural habitats for many plants and animals throughout the world and also provide important protection for water supplies and quality. Global climate change can lead to altered weather patterns, droughts, forest fires and the spread of plant pests and diseases.

Plants, animals and ecosystems

Global climate change is affecting patterns of weather and vegetation. Most plants and animals need very specific climatic conditions to thrive and survive. Plants and animals can adapt to slow changes in climate, but sudden changes can affect the ability of organisms to survive in particular locations. Climate change can also affect the life cycles/patterns of both plants and animals. For example, as temperatures increase, plants may bloom earlier in the spring and survive longer into the summer.

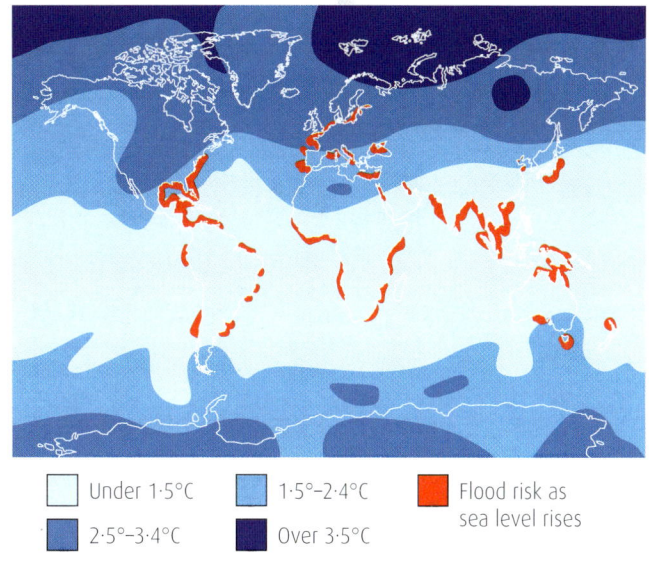

☐ Under 1·5°C	☐ 1·5°–2·4°C	☐ Flood risk as sea level rises
☐ 2·5°–3·4°C	☐ Over 3·5°C	

Areas at risk of flooding as a result of increasing global temperatures.

Coral reefs

Warming ocean temperatures have led to the bleaching of coral reefs. Rising sea temperatures can kill the colourful algae necessary for the health and survival of corals. Coral reefs are therefore under threat and are becoming rarer in tropical and subtropical waters. Loss of coral reefs affects the habitat of many aquatic species and the natural cycles of the oceans.

Drought.

Rising temperatures can lead to an increase in tropical storms.

Extreme weather conditions

Since the last ice age, which ended about 11 000 years ago, the temperature of the Earth has been relatively stable at about 14°C; however, this average temperature has been increasing in recent decades and this has affected global weather patterns. Scientists believe that extreme weather (for example, heat waves, tropical storms, flooding and droughts) is directly related to increasing global temperatures.

Altered habitats

Changes in rainfall and temperature affect the habitats of plants and animals. The following changes have been reported:

- Spruce bark beetles have boomed in Alaska as a result of 20 years of warmer summers – these beetles have chewed up just over 1.6 million hectares of spruce trees.
- Some butterflies, foxes and alpine plants have migrated further north or to higher, cooler areas.
- Numbers of Adélie penguins in Antarctica have decreased from 32 000 to 11 000 breeding pairs in 30 years.
- The first documented species extinction as a result of climate change was the death of the last golden toad in Central America in 1999.
- Polar bears may be extinct in as little as 100 years as a result of melting polar ice.
- Several states in the USA have already lost their official birds as they move to cooler climates, including the Baltimore oriole of Maryland, the black-capped chickadee of Massachusetts and the American goldfinch of Iowa.
- Species that rely on one another are no longer synchronised, for example, plants are blooming before the insects needed to pollinate them are flying.

This species of golden toad is now classified as extinct.

🔵 DON'T FORGET

Weather predictions and forecasting may become more difficult as the effects of climate change are unpredictable. In the UK, we have already seen examples of unseasonal cold spells and storms in recent years.

🔵 DON'T FORGET

You may have other examples from your school – use the ones you feel most comfortable with!

✔️ ONLINE TEST

Head to www.brightredbooks.net to test yourself on the effects of climate change.

 ## THINGS TO DO AND THINK ABOUT

Summarise the global effects of climate change and account for these changes.

LOCAL AND GLOBAL EFFECTS OF CLIMATE CHANGE 2

ONLINE

Learn more about the economic impact of climate change by following the link at www.brightredbooks.net

ECONOMIC EFFECTS OF CLIMATE CHANGE

Reduced development

Climate change is affecting the global economy. If global carbon emissions are not reduced dramatically, then many countries will find that the cost of controlling this issue will directly affect their GDP and, consequently, their development.

Reduced water supplies

Warmer temperatures are affecting global water supplies as rainfall patterns vary. This means that there is less water available for agriculture and food production. This is a big concern, particularly where populations are growing and land degradation is taking hold, for example, in the Sahel region. This will lead to a decrease in crop yields and an increased need for further irrigation, which will become both difficult and increasingly expensive.

A reduction in snowfall will affect ski resorts.

Tourism

Tourism is likely to be affected, particularly in mountainous ski resorts. Economies that rely on skiing as a form of income suffer as the ski season is reduced or disappears through a lack of snow. Increased temperatures in other areas, for example, the Mediterranean, will contribute to an increased rate of desertification, making them unsustainable as tourist destinations.

Increasing desertification may affect the sustainability of Mediterranean resorts.

SOCIAL EFFECTS OF CLIMATE CHANGE

Coastal cities will be more affected by flooding.

Migration

When conditions become harsh or unbearable, people will migrate to other areas. People tend to move away from areas of drought as they cannot irrigate their land to grow crops or find fertile ground for their animals to graze on.

Flooding also forces people to migrate. In Bangladesh, 17 million people live with the threat of flooding. This is as a result of an increasing population and increasing numbers of people living in coastal cities affected by flooding.

contd

Health

An increased risk of flooding from rising sea levels and intense storms will increase the risk of water-borne diseases such as malaria. Increases in temperature and rising water levels are ideal breeding grounds for mosquitoes. Diseases will spread more easily as the rising population causes unsanitary and overcrowded conditions.

Increases in temperature create ideal conditions for bacterial diseases to spread, which will affect both death and infant mortality rates in many countries.

Heat waves can cause illness and death in young and elderly people. The heat wave that hit the UK in 2003 is estimated to have caused over 2000 extra deaths; it caused a further 35 000 deaths in mainland Europe.

Water storage containers are ideal breeding grounds for mosquitoes.

ADVANTAGES OF CLIMATE CHANGE

Although climate change mainly brings about negative effects, there are also some positive effects of the changes in global temperatures and varying rainfall patterns:

- **Vegetation** – the Canadian Prairies – an existing major area for wheat growth – with warmer temperatures, new fruits and crops may be grown, along with an overall increase in the wheat production area.

- **Oceans** – increased fish stock in many oceans.

- **Landscape** – the summer season has been extended, which means increased revenue for some tourist destinations.

- **Hydrology** – increased awareness and education about water conservation means that less water is wasted globally.

- **Population** – densely populated areas are being dispersed due to flooding – this means that there is less pressure on this land.

THINGS TO DO AND THINK ABOUT

Describe the positive effects of climate change. Give reasons why some scientists may describe it as having positive consequences on the global economy.

In the exam, you will need to know the following for the Climate Change topic:
- physical and human causes
- local and global effects
- management strategies and their limitations

 DON'T FORGET

For more on this, read the section on malaria, pp. 82–83.

DON'T FORGET

There are positive and negative effects of global climate change. You should have examples of each to give in the exam.

ONLINE TEST

Head to www.brightredbooks.net to test yourself on the effects of climate change.

INDEX